SpringerBriefs in Applied Sciences and Technology

For further volumes:
http://www.springer.com/series/8884

Anna Bernstad Saraiva Schott
Henrik Aspegren · Mimmi Bissmont
Jes la Cour Jansen

Modern Solid Waste Management in Practice

The City of Malmö Experience

 Springer

Anna Bernstad Saraiva Schott
Jes la Cour Jansen
Lund University
Lund
Sweden

Henrik Aspegren
Mimmi Bissmont
VA SYD
Malmö
Sweden

ISSN 2191-530X
ISSN 2191-5318 (electronic)
ISBN 978-1-4471-6262-9
ISBN 978-1-4471-6263-6 (eBook)
DOI 10.1007/978-1-4471-6263-6
Springer London Heidelberg New York Dordrecht

Library of Congress Control Number: 2013954024

Printed on acid-free paper

Springer is part of Springer Science+Business Media (www.springer.com)

Preface

This book provides an example of how municipalities can work toward increased sustainability within the solid waste management area through an overview of the development in the City of Malmö, Sweden over the last decade. Organizational structures, collaboration forms with industry and academia, technologies for waste collection and treatment, as well as methods for evaluation developed in the city, are presented, based primarily on two case studies performed from 2000 to 2010. The book also discusses future challenges in the search for constantly increasing sustainability in the urban solid waste management area.

The book contains both an overall description of the case of the City of Malmö, presentation of research results based on several case studies performed in collaboration between the City of Malmö, private industry and the academia, as well as presentation of methods that can be used for evaluation of the sustainability of solid waste management strategies, which can also be transferred and used in other cities and contexts. Thus, the book is aimed at public servants and policy-makers at municipal/regional level, research and development-oriented waste management industries, as well as university professors and students.

Acknowledgments

Stig Edner, Sysav Utveckling
Åse Dannestam, MKB fastigheter
Savita Upadhyaya, VA SYD
Britt-Marie Fagerström, Tyréns
Tor Fossum, Malmö City Planning Office
Lotta Hansson, Malmö City Planning Office
Marie Castor, SWECO
Anna Granberg
Åsa Davidsson, Water and Environmental Engineering, Lund University
Language review: Brian Hazlehurst (BrianHaz@wb.com.br)

Contents

Abstract

Sustainability has, since the end of the 1990s, been a leading word in the solid waste management in the City of Malmö, Sweden. This book presents the work performed within the city based mainly on the two full-scale experiments Bo01—novel technologies in new developments, and Augustenborg—introduction of modern waste management in existing residential areas. In both cases, a close collaboration with academia has led to thorough evaluation of used waste management strategies and technologies, and thus decision-making based on scientific outcomes. This book presents the development of organizational structures, collaboration forms among the public sector, industry and academia, technologies for waste collection and treatment as well as methods for evaluation. The book also discusses future challenges in relation to, amongst others, urban planning, decision-making structures, behavioral changes and development of waste management infrastructure, seeking constantly increased sustainability in the urban solid waste management area. Finally, the book highlights the importance of viewing solid waste management as a core element in the development of the sustainable city.

Keywords Sustainable urban development · Solid waste management · Organic waste · Food waste grinders · Vacuum systems · Waste composition analysis · Life-cycle assessment · Triple helix

Chapter 1
Sustainable Waste Management in a Changing Environment

Keywords Solid waste management · Modern municipal waste management · Solid waste legislation · Environmental objectives · Demography · EU waste hierarchy

1.1 The City of Malmö: Aiming to Become the World Leader in Sustainable Solid Waste Management

World class waste management is a prerequisite for sustainable urban development.

This has been the overarching vision for the development of solid waste management in Malmö over the last decade. The current municipal waste management plan has a deeply holistic view, acknowledging the strong link between sustainable consumption and waste management, and giving decision makers in the municipal waste management organization a strong mandate to prioritize environmentally beneficial alternatives over less costly but more polluting ones. This vision also acknowledges the need to incorporate waste management planning into the wider sustainable urban development context—not as an "add-on," but as part of the core in any development project.

But how did the City of Malmö reach this point? The aim of this book is to describe the trajectory over the decade, including strategic thinking, organizational developments, technology choices, and evaluation methods. One of the main aims is also to describe the triple helix collaboration form used in the City of Malmö, where the municipality has developed strategies and achieved high performance in the area of solid waste management together with both the private sector and academia.

A. Bernstad Saraiva Schott et al., *Modern Solid Waste Management in Practice*, SpringerBriefs in Applied Sciences and Technology, DOI: 10.1007/978-1-4471-6263-6_1, © The Author(s) 2013

The development within the solid waste management area in the Malmö over decade has resulted in five major outcomes:

- Development of forms and methods for successful triple helix collaboration in full-scale development projects.
- Applying multidisciplinary evaluation methods for assessment of different aspects of solid waste management systems.
- Using life-cycle assessment methodology as a decision support tool for waste collection and treatment alternatives, as well as for identification of hot spots with high environmental impact on the solid waste management treatment chain as a basis for continuous improvements.
- Increasing the view of solid waste management as one of the core areas in urban development.
- Develop new and improve known systems and technologies for urban solid waste management.

Following this, the process of reaching these outcomes is described and discussed, based on four full-scale development projects performed in the City of Malmö over the period 1999–2012. While the first two projects have been finalized, the latter two are still in a startup or development phase:

- The Bo01 project—Novel technologies in new developments.
- The Augustenborg project—Introducing modern waste management in existing residential areas.
- Fullriggaren—Building on previous experiences for further improvements.
- Hyllie allé—Achieving integration of solid waste management in the urban planning process.

But first, the context in which these developments have been made must be described. Thus, in this chapter, the outline of Swedish waste management is presented in relation to national and regional legislation and objectives. In Chap. 2, the history of Malmö over the last few decades is presented in relation to its demographic and economic developments. Also, the waste management organization in the city is presented. Chapter 3 gives an introduction to the collaborative structures developed between academia, the municipality and industry in the area of solid waste management in Malmö over the last few decades, while Chap. 4 presents the two development projects, Bo01 and Augustenborg, in further detail, including the methods used for evaluation of the projects. In Chap. 5, the outcomes of the projects are presented and discussed in the framework of sustainable solid waste management. Chapter 6 provides an overview of the two new development projects, Fullriggaren and Hyllie allé, and discusses future challenges related to solid waste management in general and for Malmö in particular.

1.2 Waste Management then and Now

According to the UNEP (2013), some 11.2 billion metric tons of solid waste are currently being collected around the world every year. The decay of the organic fraction in this waste is contributing to around 5% of the global greenhouse gas emissions. Increased focus on waste management in development work, an increased focus on material recycling, and other means for reduction of negative environmental impacts from waste management, may have to some extent, decreased the negative impact of solid waste management over the last few decades (UN-HABITAT 2010). Examples of new legislation in the area of solid waste management have been seen in areas, such as the EU (EU waste framework directive 2008), Brazil (Brazilian Parliament 2010) and South Africa (Republic of South Africa 2008) with an enforced focus on minimizing the negative environmental impacts of solid waste management. However, the last few decades have also shown rising amounts of waste generation in the world. For example, between the years 1980 and 2005, the total quantity of municipal waste per capita increased by 29 % in North America, 35 % in OECD countries (Sjöström and Östblom 2010). As highlighted by Sjöström and Östblom (2010), the same period showed a strong link between economic growth and increased generation of solid waste. Although the need for decoupling these has been discussed, examples of such are few—if any. Thus, the benefits derived from reduced impacts of solid waste management are counteracted by increased generation of waste in absolute terms.

Sweden has in many ways been progressive in development of solid waste management. A short introduction to the current situation is presented below:

A national legislation on producer responsibility was introduced in Sweden in 1993 for packaging made from glass and cardboard. This was followed in 1994 by similar regulations also for packaging made from metal, plastic, and paper, as well as newsprint. According to this legislation (Ordinance of Producers' responsibilities for packaging) all companies that produce, import, or sell packaged goods on the Swedish market shall:

- ensure that a collection system exists, through which customers and other end consumers can return used packaging
- ensure that customers receive the information they need about the collection of used packaging
- ensure that collected packaging is recovered, recycled, and put to good use as either new raw material or energy.

Companies affected by this law have created a common system for the collection and recycling of packaging under the Ordinance. Companies affiliated with this system pay a fee in relation to the amount of packaging material their operations generate. Fees are used to cover costs for collection, transportation, and information related to the packaging recycling system. Currently, nearly 10,000 companies are affiliated with the system (FTI 2013).

Also, according to the legislation, households are required to separate material under the producer responsibility legislation from residual waste, and make use of the systems developed for waste recycling.

Since 1994, producer responsibility legislations and systems have also been implemented for cars, tires, batteries, pharmaceuticals, waste electronic equipment (e-waste), and radioactive equipment. A voluntary system has also been in place for office paper since 1996.

For waste not affected by the producer responsibility legislation, i.e., residual waste (including food waste), bulky waste, and hazardous waste,[1] collection and treatment is governed by a municipal monopoly. However, both collection and treatment can be performed by private companies, contracted by the municipality.

Collection and transportation of waste under the municipal monopoly is in almost 75 % of the Swedish municipalities, carried out by private entrepreneurs, and, in the other 25 %, by the municipality itself (Swedish Waste Management Association 2010). In the case of packaging material and newspaper, different companies can be contracted for collection of different waste fractions (i.e., cardboard, glass, plastics, etc.).

In 2005, a national environmental objective was introduced by the Swedish government, stating that, by the year 2010, 35 % of the food waste from households, industrial kitchens, and restaurants should be collected separately and treated biologically. This target was not met; in 2010, only 11 % of the food waste from these entities was collected separately for biological treatment. In 2012, the objectives were sharpened, and now state that, by 2018, 50 % of the food waste from these entities should be collected separately (SEPA 2012). By April 2012, 105 out of a total of 290 Swedish municipalities had introduced separate collection household food waste, and a further three reported that they would implement such schemes in the coming year.

In relation to information, municipalities are responsible for providing households with the necessary correct information regarding management of household waste, including waste under producer responsibility legislation. In the case of batteries and e-waste, producers have full responsibility to provide the respective information to households.

As a consequence of a more widespread use of on-site separation of recyclables[2] in Sweden in later years, Swedish real estate owners also play an increasingly important role in the management of solid household waste. This often means that real estate owners themselves also provide information regarding separate collection of packaging and newspapers, as well as residuals and food waste, to their customers (householders).

---

[1] Hazardous waste, with the exception of batteries, since 1st January 2009.

[2] Possibilities for household waste separation in direct connection to residential buildings, often in waste disposal areas inside multifamily dwellings or recycling buildings in larger residential areas.

Plants for recycling of packaging materials and newspapers are, in many cases, privately owned, whereas treatment of residual waste and bulk waste is often conducted by treatment enterprises owned by one or several municipalities. In the case of food waste, treatment plants are commonly owned by one or several municipalities, but several privately owned plants for anaerobic digestion of such waste exist.

Based on the above, it is seen that the responsibility for solid household waste management in Sweden is divided among several different agents (Fig. 1.1).

Thus, the Swedish model for management of solid household waste involves many different actors. This has resulted in several collaborative organizations, where different actors work together to fulfill policies and regulations. One example is El-retur, a collaboration between Swedish municipalities and the organization of retailers of electronic devices on the Swedish market (El-kretsen), with the objective of facilitating communication and collaboration between the municipalities and industry in order to fulfill the objectives stated in the WEEE-directive (European parliament 2002).

However, this can also result in situations where the division of responsibilities among different actors involved in the waste treatment chain not always is clear, confusion among the user of the system—households, enterprises, and the public, as well as economically and environmentally suboptimal waste management. An example of this was pointed out by Dahlén (2008); when the separate collection

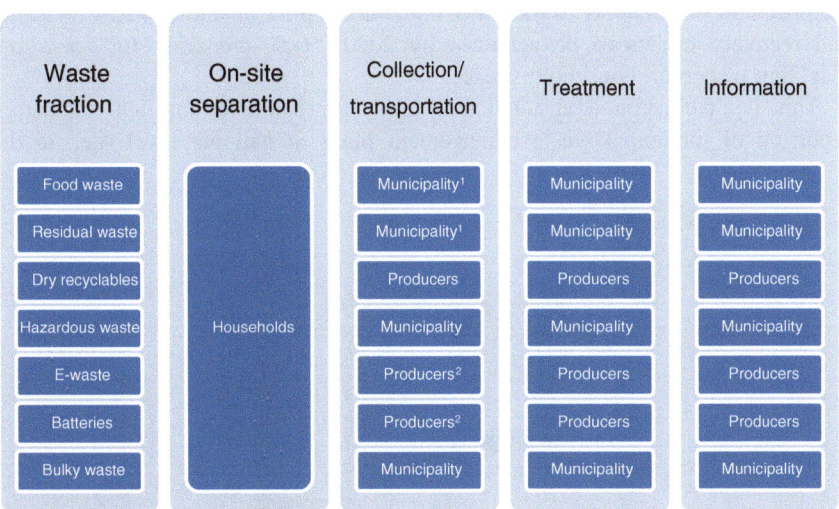

Waste fraction	On-site separation	Collection/ transportation	Treatment	Information
Food waste		Municipality[1]	Municipality	Municipality
Residual waste		Municipality[1]	Municipality	Municipality
Dry recyclables		Producers	Producers	Producers
Hazardous waste	Households	Municipality	Municipality	Municipality
E-waste		Producers[2]	Producers	Producers
Batteries		Producers[2]	Producers	Producers
Bulky waste		Municipality	Municipality	Municipality

[1] Often through private entrepreneurs.
[2] Municipalities collect e-waste and batteries through on-site collection or recycling centers commissioned by El-kretsen, which is the national co-ordinating organization for producers in order to fulfill the producer responsibility legislation on e-waste and batteries.

Fig. 1.1 Graphical representation of the division of responsibilities of Swedish solid waste management

and recycling system of the producers do not meet requirements (for example, result in overfull or untidy recycling points), usually local authorities must step in and solve the situation (Dahlén 2008). This naturally results in costs for the local waste authority—costs that must be covered by the charges collected from the inhabitants. This is also the case when recyclables under the producer responsibility legislation is collected on-site, as collection fees in such cases are paid by the property owner, but commonly transferred to households. Thus, the inhabitants pay twice: first, as a recycling-fee paid to the producers (included in the price when buying a product) and, second, as an unnecessarily high charge for solid waste management paid to the local authorities, or as an extra charge paid by households in case of on-site separate collection of packaging material.

1.3 Factors to be Considered in Modern Municipal Waste Management Policy-Making

1.3.1 Legislation and Objectives

Swedish solid waste management is largely influenced by EU legislation. The EU waste framework directive (WFD) (European Parliament 2008) was implemented in Sweden in 2011. The EU WFD emphasizes that legislation and policy of EU member states shall apply to the waste hierarchy (Fig. 1.2), and contains recycling and recovery targets to be achieved by 2020: 50 % preparing for reuse and recycling of certain household waste.

This is demonstrated at municipal level, for example, through mandatory reporting of municipal waste management plans at national level (i.e., to the

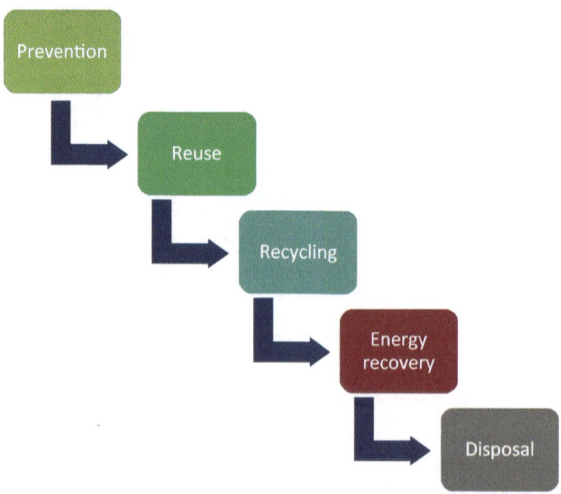

Fig. 1.2 EU waste hierarchy

Swedish Environmental Protection Agency), providing an infrastructure that makes it possible to achieve the waste recycling objectives and an enforced work related to waste minimization.

The national legislation contains, among others, responsibilities and regulations related to the transportation of household waste, hazardous household waste, e-waste, packaging, and other waste types under the producer responsibility legislation. In general, the municipalities are responsible for collection and transportation of household waste (including hazardous waste and bulk waste), other than waste fractions under producer responsibility legislation.

Since 2005, 16 National Environmental Quality Objectives should have an influence on all Swedish policy-making (Swedish Government 2004). Several objectives are related to solid waste management (Table 1.1). These objectives should be guiding both national policy-making in the waste management area, as well as municipal waste management plans.

Solid waste management is, of course, more relevant in relation to some of these objectives than others;

In the objective, "A Good Built Environment," it is stated that:

- The amount of solid waste generated in Sweden, as well as its hazardousness, should decrease.
- Fifty percent of food waste from households, industrial kitchens, and restaurants should be collected separately and biologically treated by 2018. Of this, 40 % should be treated so that its nutrient and energy content is recovered.
- The objective also sets targets for material recycling of packaging made from metal, glass, plastic, and paper/cardboard, as well as of newspapers/magazines.
- A target was established in relation to construction and demolition (C&D) waste, where 70 % of the mass of nonhazardous C&D waste should be recovered by 2020.

Table 1.1 Sweden's 16 environmental quality objectives (Swedish Government 2004)

1. Reduced climate impact
2. Clean air
3. Natural acidification only
4. A nontoxic environment
5. A protective ozone layer
6. A safe radiation environment
7. Zero eutrophication
8. Flourishing lakes and streams
9. Good-quality groundwater
10. A balanced marine environment, flourishing coastal areas, and archipelagos
11. Thriving wetlands
12. Sustainable forests
13. A varied agricultural landscape
14. A magnificent mountain landscape
15. A good built environment
16. A rich diversity of plant and animal life

Solid waste management is also relevant in relation to the objective, "Reduced climate impact," through production of renewable energy from organic waste, nutrient recovery from biologically degradable waste, increased material recycling, energy recovery from non-recyclable waste, and decreased land-filling. The same aspects also have a strong link to the objective, "Clean Air," which is also influenced by the legislation on waste incineration. Increased care in the management of waste with a high content of ozone depleting gases, as well as the introduction of systems for the collection of other types of hazardous waste, have had great importance in relation to the objectives, "A Non-Toxic Environment," "A Protective Ozone Layer," and "A Safe Radiation Environment." These systems, as well as care in relation to land-filling with waste or residues from waste incineration and destruction of hazardous waste, are of importance in relation to the objective, "Good-Quality Groundwater." Thus, solid waste management is of great importance in relation to a large part of the Swedish national environmental objectives (Fig. 1.3).

Environmental protection legislation can be relevant in relation to municipal waste management through tolerance limits regarding the noise of waste collection vehicles. Developments in working environment legislation have in recent years has improved the working conditions for waste collectors, which has resulted in legislation concerning maximum distances for manual transportation of wastebins, etc. Legislation in the traffic safety area can cause challenges for the collection of solid waste from residential areas where heavy traffic has been prohibited.

Fig. 1.3 Swedish national environmental objectives related to the objective, "Good Built Environment." *Illustration* Tobias Flygar

Reduced Climate impact

Clean Air

Safe Radiation Environment

Good Built Environment

Good Quality Groundwater

Non-Toxic Environment

Protective Ozone Layer

1.3.2 Demographic Trends, Attitudes, and Economy

Sweden, like many other countries, has, in later years, experienced a strong urbanization trend. As an example, the populations in Stockholm, Gothenburg, and Malmö (the three largest cities), in all cases, increased 8–10 % between 2005 and 2010. This has caused an increased need for urban densification, in order to avoid urban sprawl. Newly constructed areas with narrow streets and penthouse constructions on top of already existing buildings are some of the methods undertaken. However, these strategies also increase the challenges faced by solid waste management.

Not only the total population, but also the ethnic composition has changed in several Swedish municipalities. In 2010, the fraction of inhabitants born outside Sweden reached 22 % in Gothenburg and Stockholm. In Malmö, this fraction was over 30 % in the same year. The multitude of different ethnic backgrounds introduces a cultural richness in the city, but can, with regard to, solid waste management, also result in challenges, especially in relation to information strategies. Information written in Swedish might be insufficient to reach large groups. Thus, new information strategies must be used to reach households regarding separate waste collection and the environmental benefits related to this. The ethnic diversity can also result in a high diversity of gastronomic traditions, where waste types previously not common in the society, as well as other lifestyle differences, can lead to new challenges for the municipal waste management system.

Households and enterprises included by the municipal monopoly of waste collection and treatment have expectations of sound, noncomplicated, effective, hygienic, and cost-efficient solid waste management, which must be addressed by the municipality in order to maintain the credibility of these users. This is also related to an increasing awareness of environmental issues among the users. An example of this is the questioning of the environmental gains related to separate collection and recycling of waste, when this can be connected to increased need for transportation, washing of packaging material prior to source-segregation, etc.

Today, several municipalities are also working to constantly increase esthetic values in the urban environment. Challenges can emerge in relation to such values and the need for public recycling points, public litter bins, buildings for on-site separate collection in residential areas, etc. Source segregation will commonly also increase the space needed for waste disposal. Thus, introducing for on-site separate collection in existing buildings and areas with dense construction, such as older city centers, is often challenging.

Municipalities are challenged by increased demands from users to be cost-effective, i.e., provide good service at low cost. According to Swedish legislation, all costs connected to municipal waste management shall be covered by a charge paid by the users of the waste management services, rather than through taxes. Also, charges must be set on levels covering only the actual costs of the waste management, without generating profits (SFS 1991).

The charge consists of two parts; a fixed charge, covering costs of public recycling centers, transportation, and treatment of bulky waste and hazardous waste,[3] planning and administration of waste management as well as information; and another, dynamic charge, covering costs for transportation, and treatment of residual waste and food waste.

Several municipalities have the ambition to set their waste management charges in a way that promotes environmentally friendly behavior among users. One example is waste-based charges on residual waste or a decreased charge related to collection and treatment of separately collected food waste, at the expense of a higher charge on residual waste. However, as presented above, the municipality's possibilities to adjust waste management charges in a sense which promotes environmental procurement are limited by Swedish legislation.

The cost of waste management is, however, generally low in comparison to many other costs borne by households. Several studies have shown that the impact from increased municipal waste management charges on the household's waste behavior is low, especially in rental areas (Swedish Waste Management Association 2011). However, a differentiation of charges for different types of waste collection subscriptions (for example, related to collection frequency) can be of importance in relation to detached houses where the cost is more visible to the household (Bernstad 2013).

1.3.3 Potential for Conflicting Interests

The text above exemplifies several aspects of how solid waste management in modern cities has to address a series of different requirements imposed on different actors at the local, national, and supranational levels (Fig. 1.4). In this mosaic of actors and interests, certain interests might sometimes conflict. Some examples:

- Legislation of public procurement might make it difficult for the municipality to be flexible in relation to special collection and treatment arrangements in cases where this could be motivated from an environmental point of view and in line with national environmental objectives.
- Increased competition for specific type of wastes in an open market (for example, organic waste with high biogas potential) could make it difficult for the municipality to plan investments in treatment technology, although this would be in line with national environmental objectives.
- Large investments in incineration plants and district heating systems might decrease the interest in increased on-site separation of recyclables with high energy content, such as plastics.

---

[3] Costs related to collection of recyclable materials under the producer responsibility ordinances should be covered by producers.

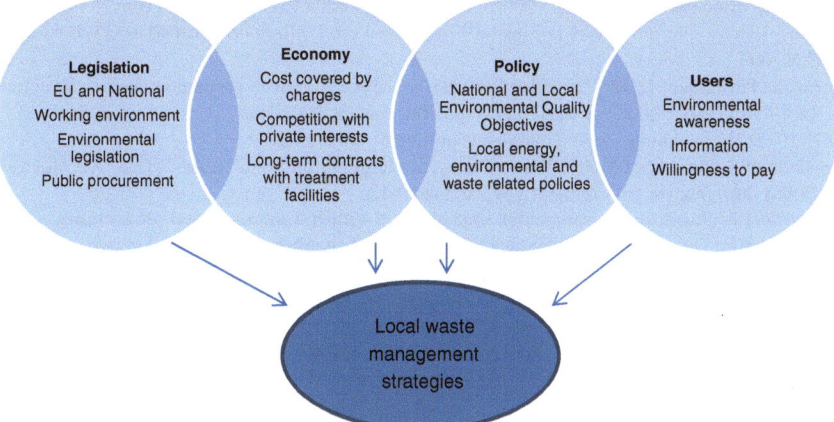

Fig. 1.4 Graphical representations of different interests affecting municipal policy-making in modern cities, based on the example of Malmö

- Legislation related to the working environment might reduce possibilities for separate collection of heavy waste fractions, such as food waste or hazardous waste.
- Improving possibilities of on-site separation of dry recyclables can be connected to increased costs for packaging producers (through the producer legislation on packaging), while decreasing the municipality's costs of handling residual waste.
- Users of municipal waste management services (i.e., mostly households) might not have a high willingness to pay for waste management services. At the same time, the municipality must abide by the legislation and achieve objectives with a budget covered by user charges.

Balancing different interests and finding solutions where all the actors involved are satisfied is therefore one of the greatest challenges in modern municipal waste management policy-making and the daily work of the municipal waste management department.

References

Bernstad (2013) Bakgrundsstudie inom projektet: KOMKOM – Kommunal kommunikations-strategi för ökad och förbättrad insamling av matavfall, Biogas Syd, Malmö (in Swedish)

Brazilian Parliament (2010) National solid waste law of August 2nd, 2010. Law 12.305/2010. Brasilia

Dahlén L (2008) Household waste collection: factors and variations, Doctoral thesis Luleå University of Technology, p 33

European Parliament (2002) Directive 2002/96/EC of the European parliament and of the council of 27 January 2003 on waste electrical and electronic equipment (WEEE)

EU waste framework directive (2008) Directive 2008/98/EC on waste (Waste Framework Directive). http://eur-lex.europa.eu/LexUriServ/LexUriServ.do?uri=CELEX:32008L0098: EN:NOT

European Parliament (2008) Directive 2008/98/EC of the European parliament and of the council of 19 November 2008 on waste and repealing certain directives

FTI (2013) Förpacknings- och Tidningsinsamlingen. www.ftiab.se

Republic of South Africa (2008) Waste Policy and Regulation, WASTE Act, 2008 (Act No. 59 of 2008). http://sawic.environment.gov.za/?menu=13

SEPA (2012) Swedish environmental objectives. Swedish Environmental Protection Agency, Stockholm.   http://www.naturvardsverket.se/Miljoarbete-i-samhallet/Sveriges-miljomal/ Miljokvalitetsmalen/

SFS (1991) Kommunallagen, Svensk Författningssamling (SFS) 1991:900

Sjöström M, Östblom G (2010) Decoupling waste generation from economic growth. A CGE analysis of the Swedish case. Ecol Econ 69:1545–1552

Swedish Government (2004) Environmental quality objectives – a shared responsibility, 2004/ 05:150, Stockholm

Swedish Waste Management Association (2010) Swedish waste management 2009. Swedish Waste Management Association, Malmö

Swedish Waste Management Association (2011) Viktbaserad avfallstaxa – en litteraturöversikt/ Weight based fees – a literature review. Report U2011:10

UN-HABITAT (2010) Solid waste management in the world's cities : water and sanitation in the world's cities 2010. UN-HABITAT, Nairobi. ISSN: 9781849711708

UNEP (2013) Climate change mitigation, waste, http://www.unep.org/climatechange/mitigation/ Waste/tabid/104349/Default.aspx

Chapter 2
The City of Malmö as a Case Study

Keywords Urban development · Local waste management · Waste management organization · Urban transition

2.1 From Industries to Knowledge Production

Malmö is traditionally known as an industrial city. The main economic activity in the city from the beginning of the nineteenth century to the end of the twentieth century was connected to the harbor and shipping industry. The 138-m tall Kockums crane was the symbol of the city and is used for the construction of more than 75 ships, but the shipbuilding industry entered a deep crisis in late 1970s, resulting in increased unemployment in Malmö and a downturn for the whole city. Many people left the city and the population decreased from 265,000 in 1970 to 224,000 in 1990. The crisis worsened through the closure of the cement and car manufacturing industry at the beginning of the 1990s, a period when Sweden as a whole passed through a deep economic recession. The early 1990s was also a period of great instability in many parts of the world. Sweden received large numbers of refugees from Somalia and the former Yugoslavia. Many of the newly arrived were placed in Malmö, as the city had a large amount of empty apartments many of which were owned by the municipality itself through the municipal housing company, Malmö Kommunala Bostäder (MKB). The immigration helped to counteract the population decrease and gave the city the multicultural identity it is well known for today.

Through the later part of the 1990s and onward, the city has struggled to improve its economic situation. An important step in this process was the establishment of Malmö University in 1998. It was built in the Western Harbor area,

A. Bernstad Saraiva Schott et al., *Modern Solid Waste Management in Practice*,
SpringerBriefs in Applied Sciences and Technology, DOI: 10.1007/978-1-4471-6263-6_2,
© The Author(s) 2013

Fig. 2.1 The Western Harbor with two landmarks in Malmö then and now: Kockum crane (*left*) and Turning Torso (*right*). *Photo* Malmö Urban Planning Office

where large parts of the former shipbuilding industry had been located (Fig. 2.1). Another important step was the construction of the Öresund-bridge. In 2000, the bridge connecting Malmö to Copenhagen was inaugurated, facilitating both personal travel and transportation of cargo between Sweden and the rest of Europe.

Malmö is today, in many ways, a unique city in Sweden—the closeness to Copenhagen and continental Europe, an increasing multiculturalism and transition from industrial city to knowledge-based city has, over the last few decades, changed the profile of the city as well as the general conception of it held by its inhabitants.

Malmö is also a young city in comparison to many other Swedish cities. In 2011, 30 % of the population was under 25, 33 % 25–44, while 15 % were over 65. The University of Malmö has attracted many students to the city. Also, as a consequence of improved possibilities for commuting, many students from Lund University reside in Malmö.

Thus, over the last few decades, Malmö has changed vastly—both in terms of its economic development, demography, and self-image. This has also had an effect on the solid waste management in the city.

2.2 Waste Management in Malmö

The Establishment for Waste Management (Malmö Renhållningsverk) was founded in 1898 as a department in the municipal structure. The municipal waste department was responsible for both planning of waste management and waste collection/transportation services. By 1992, the department was privatized, as a part of a general privatization wave in Sweden at this time. The collection/

transportation service was sold to a private entrepreneur and the planning section was developed into an ordering organization, placed under the municipal Department for Public Spaces. By 2000, the waste management planning section was transferred to the municipal Department for Water and Wastewater handling. One of the reasons for this was that both wastewater treatment services and solid waste management services are financed by fees, while this is not the case for other activities managed by the Department for Public Spaces (which are tax-financed). By 2008, the Departments for Water and Wastewater handling in Malmö and the neighbor municipality of Lund were merged, creating the municipal enterprise VA SYD. However, the city of Malmö and politicians in the Technical Committee still has the political leadership over the solid waste management sector. As the main argument for the merge of the two municipal departments was to create collaboration in the field of wastewater treatment, the future organization of solid waste management in the City of Malmö is still under discussion. Sysav (South Scania Waste Company) was created in 1974 with the mission to coordinate solid waste management in several municipalities in southern Scania. The company is owned by 14 municipalities, one of them is the City of Malmö. Thus, treatment of municipal solid waste has since then been performed through Sysav.

Thus, both the municipal mission and organization of solid waste management within the City of Malmö have changed over the years, and might also change in the future. In its current form, a Department of Waste Management within VA SYD and has the mission of planning and managing collection and treatment of municipal solid waste which not is covered by producer responsibility ordinances.

In 2011, the amount of residual, garden, and food waste collected in the City of Malmö reached 90,800 tons. Collection and treatment of waste is not performed within the organization. These services are bought from the private sector and performed by an entrepreneur.

In 2009, 75 % of the Swedish municipalities had outsourced collection and transportation of solid waste to private entrepreneurs (Swedish Waste Management Association, 2010). This is the case also in the City of Malmö. However, the municipality still has large potentials of influencing the collection and transportation in these cases. As an example, municipal enterprises are obliged to follow municipal policies related to environmental precaution in all ordering of services and products from external parties. Thus, in the case of Malmö, the municipal policy states that environmental, social, and ethical requirements should be made in all public contracts and that requirements made should aim for long-term sustainable development (City of Malmö, 2012). However, the policy also states that requirements should be proportional, which implies that a balance must be found between high environmental ambitions and an economic reality. Based on this, the municipality has in later years required that collection of waste should be performed in vehicles run on biogas. This is a strategy which does not only decrease environmental burdens related to waste collection, but also stimulates the biogas market through increasing the demand.

According to the company's internal vision, VA SYD should contribute to a sustainable development through recycling oriented waste and wastewater management. In line with this vision, several research and development projects have been initiated over the years and research and development is today integrated in the daily work within the Department for Waste Management.

References

City of Malmö (2012) Public procurement guidelines. http://www.malmo.se/Foretagare/
   Offentliga-upphandlingar/Regelverk-och-policy/Hallbar-upphandling.html
Swedish waste management association (2010) Swedish waste management 2009. Swedish
   Waste Management Association, Malmö, Sweden

Chapter 3
Collaboration with External Partners: Solid Waste Management in Development

Keywords Triple helix · Cross-institutional collaboration · Project organization · Research and development

As discussed above, local waste management in Sweden is influenced by several different factors on both national and supranational level, which during later decades has been striving toward increased sustainability. However, the local authority can chose to be more and less proactive in relation to regulations and demands. In the case of Malmö, the city chose to take a proactive role in finding new solutions for efficient and sustainable urban solid waste management. This work has, in many cases, taken the form of development projects in specific geographical areas and in collaboration with external parties, based on a triple helix model, with participants from the public, academy, and industry/private sectors, in most cases represented by facility owners or waste treatment enterprises.

The concept of triple helix has been developing since the early 1990s. The theoretical framework provides parallel and slightly different definitions of the concept. Broadly, the concept can be explained by the difference between the two conceptual models in Fig. 3.1. In the laissez-faire model, the three spheres are separated and have clearly defined and distinguished objectives; the task of the university is to perform basic research and provide human resources, the task of the industry is to create firms acting on markets to provide economic wealth for their owners, and the task of the governments is limited to addressing market failures. To the right, the three spheres have been integrated and the boundaries have been blurred (Gibbons et al. 1994). The interaction across otherwise defended boundaries is commonly, but not necessarily, mediated by organizations such as industrial liaison, technology transfer, or contract offices.

Thus, the triple helix model provides a direct communication between these three spheres. According to the theoretical framework, this provides simulation of innovation in the focal point, in which these spheres meet. This is an area where the perspectives of the different spheres encounter; common challenges and objectives can be discussed from the different perspectives of the three spheres in order to find innovative win–win solutions.

A. Bernstad Saraiva Schott et al., *Modern Solid Waste Management in Practice*, SpringerBriefs in Applied Sciences and Technology, DOI: 10.1007/978-1-4471-6263-6_3, © The Author(s) 2013

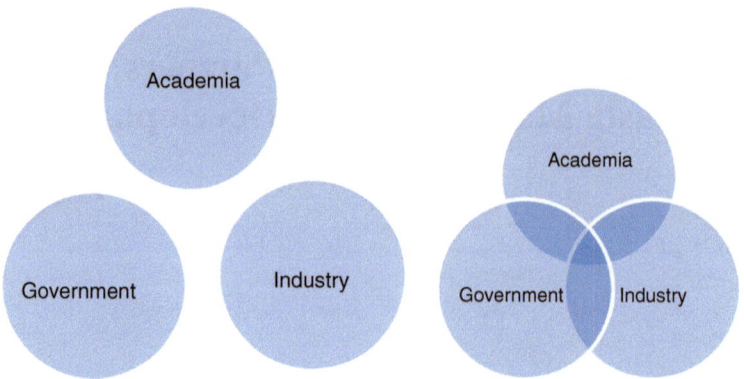

Fig. 3.1 Conceptual models of the laissez-faire model (*left*) and triple helix model (*right*)

Cultural evolution is, according to Etzkowitz and Leydesdorff (2000), driven by individuals and groups who make (1) conscious decisions as well as (2) the appearance of unintended consequences. Innovations in the framework of triple helix result from the network of relations which add surplus value on top of the objectives which have been predefined.

Another part of the theoretical framework proposed in relation to the Triple Helix concept suggests that the three institutional spheres, in addition to fulfilling their traditional functions, each also "takes the role of the other," and thus, execute new roles as well as their traditional function (Etzkowitz et al. 1998).

Development project within the area of solid waste management in Malmö has been guided by a growing view on solid waste as a resource, and special focus has been directed towards food waste management for several reasons:

- Food waste is the largest fraction of household solid waste, comprising of around 40 % of Swedish household waste.
- Separation of food waste from other waste fractions can improve qualities in residual waste for incineration purposes.
- The energy content in food waste can be used as biogas for vehicles.
- Food waste contains nutrients which can be recycled.

The concept of phosphorus recycling from organic waste had led to a previous collaboration between the City of Malmö and Lund University within the area of wastewater treatment. A number of industrial Ph.D. projects, where Ph.D. students divided their time between jobs within the municipal wastewater treatment sector and the academy, had been realized. This strategy was used to lower the borders between the academy and theory on the one side and the industry and the practice on the other. The concept had been seen as fruitful for both parties as researchers gained access to full-scale plants for their trials and the research findings were applied in practice immediately, which could improve the performance of the municipal systems.

This concept was in the Bo01 project transferred to the area of solid waste management. However, as this research area is less established, several voices—both from the side of the municipality as well as the academy, were raised questioning the aim with the collaboration. The personal contacts developed within the wastewater management area and the fact that solid waste management in this very same period was incorporated in the previous water treatment department in the municipality of Malmö, were therefore helpful in this process.

References

Etzkowitz H, Leydesdorff L (2000) The dynamics of innovation: from national systems and mode 2 to a triple helix of university–industry–government relations. Res Policy 29:109–123

Etzkowitz H, Webster A, Healey P (1998) Capitalizing knowledge: the intersection of industry and academia. State University of New York Press, Albany

Gibbons M, Limoges C, Nowotny H, Schwartzman S, Scott P, Trow M (1994) The new production of knowledge: the dynamics of science and research in contemporary societies. Sage, London

Chapter 4
From Idea to Reality: The City as a Test Bed

Keywords Development projects · Novel waste collection technology · Vacuum systems · Food waste grinders · Evaluation methods · Life-cycle assessment

Little guidance is provided for local actors in their work toward achievement of national environmental objectives related to waste management and the overall goal stated in national waste plans. Thus, there is a need for proactive and innovative measures to be taken at municipal level. In Malmö, different geographical areas of the city have been used as a test bed for innovative waste management strategies. This has provided possibilities for parallel introduction of different technologies, full-scale management systems, and information strategies, and thus comparative evaluations of these.

Both the Bo01 as well as the Augustenborg project can be described as full-scale experiments. As can be seen in Fig. 4.1, the two areas are located in different parts of Malmö. Although the two geographical areas differ greatly, the aims in these projects were similar (Table 4.1).

4.1 The Bo01 Project: Novel Technologies in New Developments

The concept of ecological sustainability, at the end of the 1990s, when the Bo01 project was initialized, was still, to a great extent, connected to eco-villages. Decentralization and small-scale systems were still regarded as the way to increase sustainability. Thus, initially, the structures in the Bo01 were planned, to a great degree, very similarly to an eco-village, aiming at self-sustainable systems. Hundred percent of the energy used in the area was to be produced in the same

Fig. 4.1 Malmö with the Bo01 and Augustenborg (dotted line) areas

Table 4.1 Aims connected to solid waste management in the areas Bo01 and Augustenborg

Bo01	Augustenborg
• Increase recycling rates of dry recyclables	• Increase recycling rates of dry recyclables
• Evaluate introduction of novel technologies for on-site separation of food waste	• Evaluate introduction of on-site separation of food waste, e-waste, hazardous waste, and fat, oil, and grease (FOG)
• Evaluate user friendliness	• Evaluate user friendliness
• Evaluate environmental benefits from introduced changes and potential improvements	• Evaluate environmental benefits from introduced changes and potential further improvements
• Decrease need for waste transportation and use of heavy vehicles in the area	• Evaluate effect of information strategy
• Including inhabitants in waste management	• Evaluate potential for food waste minimization
• Decrease amount of hazardous waste in other waste fractions	

area, and the objective for the solid waste management was to close the loops within the actual area. The solid waste management strategy for the area was, therefore, largely to be performed through decentralized waste treatment and recovery, and contained:

- Separate collection of household food waste.
- Recycling stations for packaging materials and hazardous waste strategically placed in the area to ensure minimal walking distance for residents.
- Decentralized and on-site composting of garden and park waste.
- Use of woodchips from Christmas trees and branches from garden and park waste in plantations in the area.
- Use of biogas, produced from food waste generated in the area, in waste collection vehicles used in the area.
- On-site separation of waste in public spaces, using multicompartment wastebins.
- A re-design and re-use workshop where furniture, electronic waste, and other bulky waste, which would have been disposed of as waste, is refurbished, redesigned, and sold.
- Involvement of households in the waste management at an early stage.
- Feedback to households through use of electronic keys for registration of waste disposal and display of statistics on the environmental benefits achieved through on-site separation of recyclables individually for each household.

The initial plans also contained both a biogas and a wind power plant centrally placed in the residential area. However, it was soon realized that this would not result in an attractive living environment. The wind power plant was thus re-planned to be installed in the industrial harbor, but still within sight of the Bo01 area, in order to ensure understanding of energy use and production in the area. The planned biogas plant was removed only after the performance of a systems study, comparing decentralized and centralized solutions for organic solid waste management in the area. Also, other parts of the strategy presented above were never realized due to several reasons.

The main funder of the waste management strategy was the National Environmental Protection Agency through a program for local development. However, several other areas such as local production of renewable energy, local storm water management, and sustainable transportation were prioritized, while solid waste management gained less financial support. Another important factor was that the strategy for solid waste management was developed by public servants in the municipal waste management section, which, at that point, was a division under the Department of Parks and Public Spaces, in close collaboration with the Department for Environmental Protection. However, several of the ideas were not in line with the interests of other divisions of the municipality. As an example, the suggested recycling stations were seen as conflicting with aesthetic values, which were of great importance to the Urban Planning Office, and the suggested feedback to households was seen as conflicting with individual integrity. Also, early involvement of households in waste management became difficult, as many of the apartments in the area remained unrented for an extensive time, as they were rather expensive. Third, the Bo01 area was presented as an example of the cities of tomorrow, and solid waste management became an important part of this futuristic concept. With this background, novel waste management solutions impressed decision-makers and technological development became a focal point in the developed solid waste management strategy.

Developers and property owners participating in the fair were not interested in using land with a high market value for construction of recycling buildings, and initiatives resulting in saving space became highly relevant significant. There was also a strong focus on creating a narrow dense area, which increased the need to minimize the use of heavy vehicles in the area. To compensate for the lack of visibility of structures for solid waste management in the area, it was seen as positive if the systems were visually present in the household kitchen, in line with the sustainability profile of the area.

Thus, the initial focus on small-scale solutions, closing of loops, reuse and behavioral changes, was abandoned. Instead, development and evaluation of efficient collection systems for organic solid waste became the focus in the project, and two novel systems for food waste collection were introduced in the area (Fig. 4.2).

4.1.1 Unconventional Food Waste Disposal System

Food waste disposers (FWD) have been suggested as a practical way to establish on-site separation of household food waste without increasing transportation, odor, or need for more wastebins amongst others (Marashlian and El-Fadel 2005). However, there are several questions regarding the effects of FWD connected to the conventional sewage system. Bolzonella et al. (2003) stated that FWD can cause an increased organic load in the biological step at the wastewater treatment plant (WWTP) and thereby increase the energy demand for aeration in the wastewater treatment. Others have raised problems due to increased oil and grease loads at WWTPs and risk of increased H_2S production in sewage systems, potentially resulting in corrosion of cement pipes (Nilsson et al. 1990). Also, the knowledge of the quantity and quality of ground food waste from FWD systems actually reaching WWTP is still incomplete. In previous studies, it has been seen that the removal of dissolved organic matter and proteins in wastewater during sewage transportation to WWTP can be considerable (Raunkjaer et al. 1995).

In order to exploit the advantages of FWD and at the same time avoid possible negative effects in sewage systems and WWTPs, an unconventional FWD disposal system was introduced in in kitchen sinks in approximately 60 apartments in 2001 and in a further 147 apartments in a high-rise building in 2007, in the Bo01 area. This wastewater stream was led to a settling tank (volume 2.7 m^3) divided into different sections. From the tank, supernatant is led to the WWTP, while the settled material forms a sludge at the bottom of the tank, which is collected and transported for further anaerobic biological treatment. The principle of the systems is shown in Fig. 4.3, where the truck for sludge transportation is shown together with the sewer, and the WWTP for handling the nonsettled material. Sludge collection was performed once a month. The evaluation of the system installed in 2001 was

Fig. 4.2 The Bo01 area from above during the construction of the Turning Torso. *Photo* Malmö City Urban Planning Office

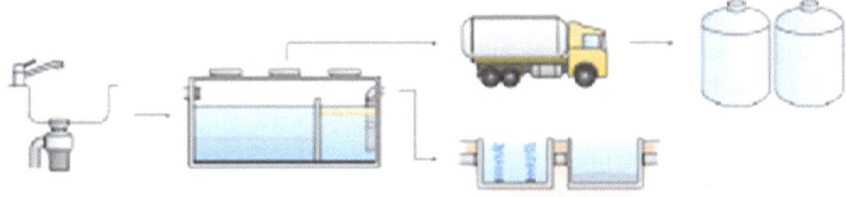

Fig. 4.3 Principles of the food waste disposer and sedimentation tank system together with the truck for transportation of the settled material, the sewage and the WWTP for handling the nonsettled material

made in 2002–2003, and the evaluation of the system installed in 2007 was made in 2010–2011.

4.1.2 Vacuum System

Five hundreds households in the Bo01 area were connected to the vacuum system, in which food waste is separately collected in paper bags, while residual waste is disposed of in nonspecific plastic bags. Households dispose of food waste and residual waste through two different inlet doors at the collection points inside the residential building or in close proximity (Fig. 4.4). From the inlet doors, the waste passes along chutes and is collected in underground tanks from which it is sucked

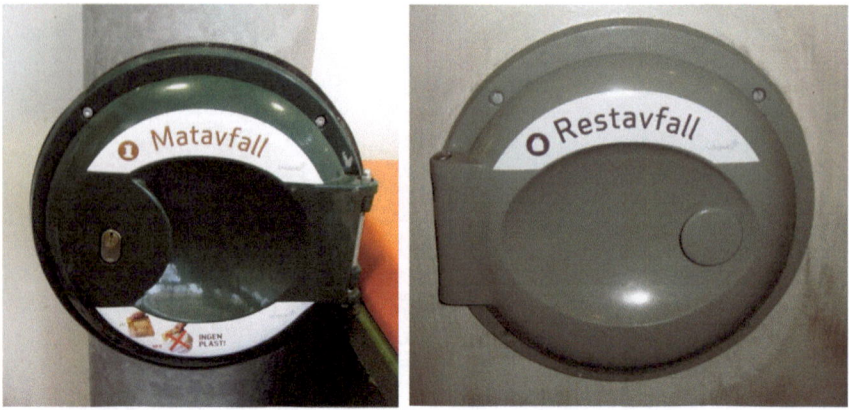

Fig. 4.4 Inlet doors for disposal of separately collected food waste in paper bags—with lock (*left*), and residual waste (*right*). *Photo* Maria Hoppe

by vacuum to a collection vehicle. The vehicle is docked to the system from an outlet in the street to which several different tanks for separately collected food waste and residual waste are connected. As collection vehicles will collect first one fraction and then the next, the same tube system can be used for both (Fig. 4.5). Since the docking point can be built at a great distance from the inlets, vacuum systems can be advantageous in densely built-up urban districts.

4.1.3 Information Strategies in the Bo01 Area

All households moving into the Bo01 area received written information in the form of a leaflet, which contained practical information regarding separate collection of food waste—i.e., what should and should not be sorted as food waste. Households were also informed on how to use FWD and practical tips on how to facilitate sorting of food waste into paper bags.

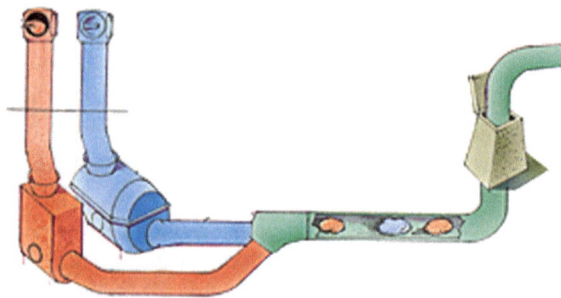

Fig. 4.5 Vacuum system with tanks and collection point for waste transportation vehicle

Several information meetings were also organized on the initiative of the Department for Waste Management shortly after households had moved into the Bo01 area, in the summer and autumn of 2001. However, no residents attended the meetings. Instead, a doorstep campaign was performed in 2002 and 2003, both among households connected to the food waste disposal system and to the vacuum system. On seven occasions, households new in the area were visited by informants from the Department for Waste Management. The primary aim of the visits was to ensure that households had received information regarding the waste disposal systems in the area. Another aim of the visit was to demonstrate the importance given by Malmö to participating in food waste sorting. Visits were, in all cases, announced in advance.

4.2 The Augustenborg Project: Introducing Modern Waste Management in Existing Residential Areas

Augustenborg is the name of a residential area in the eastern part of Malmö. The area consists of 1,631 apartments in 37 multifamily buildings. The Augustenborg project was initiated in 2008 by Malmö Municipal Housing Company (MKB). The company wanted to improve the service for environmentally beneficial waste management for their customers (households) and increase the economic transparency throughout the collection and treatment chain. They also wanted to increase knowledge regarding to what extent the efforts they made generated positive environmental benefits. Thus, a project was formed with economic resources from MKB, VA SYD and the South-Western Scania waste management company, Sysav. The overarching aim of the project was to evaluate and develop the Swedish model for solid household waste management with the focus on extended on-site separation of several different household waste fractions. In the project, current household waste streams and waste recycling behavior were to be investigated, but a clear aim in the project was also to influence the recycling behavior, both through improved possibilities of on-site recycling of new waste fractions and through use of different information strategies targeting households. The project was run in the form of a full-scale and long-term case study in the residential area, Augustenborg in Malmö, owned by MKB. Water and Environmental Engineering at the Department of Chemical Engineering at Lund University contributed to the project, assuming the task to monitor and evaluate the current status of the waste management system in the area, the effects of changes in this system taking place during the project period and potential for further improvements.

4.2.1   The History of the EcoCity Augustenborg

When built in the late 1940s, the Augustenborg residential area was a popular neighborhood, with running water and water closets indoors—which was modernity in Sweden at that time (Fig. 4.6). However, in the 1960s, several new areas with larger apartments and further from the city center—often considered as something positive at the time, were constructed in the city. Thus, by the 1970s, the flats had started to feel old-fashioned and people began moving away from the area. The many empty apartments resulted in decreased resources for necessary maintenance of buildings and outdoor spaces in the area. When Sweden received large amounts of immigrants in the early 1990s, many of them were directed to these empty apartments in the Augustenborg area.

By the end of the 1990s, the Augustenborg residential area faced social and economic problems as well as physical degradation of buildings and outdoor areas. At this point, Malmö decided that a large-scale renovation of the neighborhood was necessary. At the same time, it was decided that this should be done with an eco-logical approach, turning the area into an EcoCity. The refurbishment project was co-ordinated by Malmö and financed by the city together with MKB, the Swedish Government and the EU Life fund. Solid waste management was an important part of the project, which also contained incentives for energy efficiency, open storm water management, and support for increased urban biodiversity (Fig. 4.7).

The developments in the area had not resulted in increased costs for the inhabitants. Thus, when the Augustenborg project was initiated in 2008, the socioeconomic level in the area could be described as lower than the city average, and 15 % of the inhabitants received social welfare benefits. Also, the proportion of nonethnic inhabitants was large. According to data from 2001 to 2004, 49 % of the inhabitants were not born in Sweden, and a further 12 % were the children of

Fig. 4.6 Overview of the Augustenborg area in early 1950s

Fig. 4.7 Open storm water management system in the Augustenborg residential area. *Photo* Malmö City Urban Planning Office

parents both of whom were born outside Sweden. Several different languages were spoken in the area, Arabic, Serbo-Croatian, and Polish being the most common, after Swedish. Of the inhabitants aged 20–64 years old, 64 % had an education equivalent to upper secondary school or higher. The unemployment rate was 6 %, and 16 % of the inhabitants were over 65. The area had a turnover rate of 14–16 %, and consisted, to a great extent, of blocks of small flats, 61 % of which were one-room and 26 % two-room. The ratio of car ownership was low; 0.22 per person (in comparison with the Swedish average of 0.42 per person).

4.2.2 Infrastructure for Waste Collection in the Augustenborg Area

Since the end of the 1990s, a number of recycling buildings have been constructed in the residential area. The buildings were locked and each resident had access to one specific building by means of an electronic key. Household waste could be on-site separated into nine different fractions in the recycling buildings; glass (clear and colored), paper, plastic and metal packaging, newspapers, batteries and residual waste. Organic household waste could be disposed of in decentralized compost reactors, one in each recycling building. All in all, 13 recycling buildings were placed in the area, with the aim of reducing the distance from each household to the closest building to a maximum of 200 m (Fig. 4.8).

Thus, since 2000, households in the area have had the possibility of on-site separation of packaging and newspaper in close proximity to their home. However, little was known about the waste recycling behavior in the area. For disposal of hazardous waste, e-waste and bulky waste, tenants in the Augustenborg area were directed to the municipal waste recycling center approximately 10 km away. This

Fig. 4.8 One of 13 recycling buildings in the Augustenborg area

disposal method requires a vehicle. Statistics show that only one in five inhabitants in the area owned a car. Thus, bulky waste (including bulky e-waste) was often dumped in recycling buildings, basements or in open spaces in the area. This incorrect disposal led to vast costs for the facility owner and often resulted in time-consuming manual collection activities and uncomfortable working conditions for maintenance personal in the area. A reduction in such disposal and more controlled collection of bulky waste was, therefore, highly desirable.

Several physical changes were introduced in the area when the Augustenborg project started in 2008. Each recycling building was equipped with structures for on-site separation of electronic/electrical equipment waste (e-waste), hazardous waste, and fat, oils, and grease (FOG). Compost reactors were removed and food waste was instead to be on-site separated in paper bags for later disposal in separate wastebins in the recycling buildings (Fig. 4.9). Collected food waste is thereafter used for biogas production. An on-site collection system for bulky waste was also introduced in the area, and households could dispose of bulky waste in a mobile sorting facility, free of charge, once a month (Fig. 4.10).

For some of the above-mentioned fractions, there was little previous experience of on-site separation in Sweden. Efficient new collection and sorting systems had to be developed, while maintaining high standards of safety, as well as good accessibility and working conditions for facility managers and waste collection personnel. In the case of hazardous waste, a cabinet with a flexible one-way barrier was designed so that the boxes (and thus disposed hazardous waste) could not be removed by tenants. The locked cabinets could only be opened by facility managers, who controlled and sorted disposed hazardous waste on a weekly basis. Cabinets were emptied by the Municipal Waste Management Company when full. The cabinet is also used for on-site separation of light and low-energy bulbs.

Fig. 4.9 Interior of recycling building in the Augustenborg area after introduced changes; Overview (*left*) and bins for separately collected food waste (*right*)

Fig. 4.10 Mobile on-site collection system for bulky waste

E-waste is disposed of in open metal cages (volume 1.5 m³), which are emptied by the municipality when full (Fig. 4.11).

Households were also asked to dispose of used FOG in single-use 1.75 l plastic jars with plastic locks. Graphical instructions on how FOG should be disposed of were printed on the lock. The jars were disposed of in open boxes in the recycling buildings. Households could collect new jars from the recycling building at any time to continue the on-site separation of FOG (Fig. 4.12).

4.2.3 Information Aimed at Households

In Sweden, it is the responsibility of the municipality to provide households with correct and necessary information regarding management of household waste. However, in the case of some waste fractions, e.g., e-waste and batteries, producers

Fig. 4.11 Cabinet for collection of hazardous waste (left), and cages for collection of e-waste (right)

bear full responsibility to provide information to households. Also, it is not uncommon for treatment facilities, as well as facility owners (in the case of rental housing areas) to distribute information to households. Different agents can have different interests in relation to on-site separation of household food waste, which can be illustrated in a situation where collection and treatment of separately collected food waste is cheaper compared to handling residual waste. The agent paying for the handling has economic incentives to include as many types of on-site-separable waste as possible. In practice, this can mean that households are advised to separate flowers, soil, coffee grounds, etc., as food waste. In the case of the receiving facility, the interest can be more related to the impact from on-site separated waste on the receiving facility. Thus, households can commonly feel confused regarding what actually is right in relation to their waste sorting (Ewert et al. 2009).

Fig. 4.12 Plastic jars developed for collection of fat, oil, and grease (FOG) from households

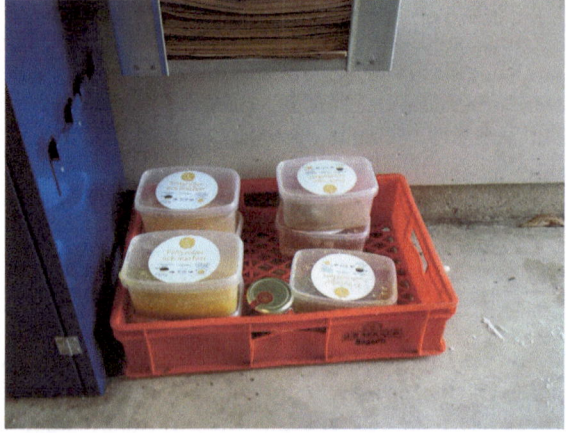

Within the Augustenborg project, an overarching information strategy was developed in collaboration among the municipality (VA SYD), the waste treatment agent (Sysav) and the facility owner (MKB). The message sent to households regarding sorting instructions was discussed within the project group, potential controversies were resolved in advance and the information gained from households was coherent. Households were also given a single contact person to whom they could direct any questions.

4.3 Methods for Evaluation

4.3.1 Evaluation of Waste Generation and Recycling Behavior

Several different methods can be used for evaluation of solid waste generation and household recycling behavior. They are all connected to different benefits and drawbacks.

- Weighing waste during collection.
  Modern waste collection vehicles can be equipped with scales in order to record weights from individual wastebins. Marking each wastebin with an individual code makes it possible to follow the development of different waste fractions generated over time. Care should be taken so that the weight of the wastebin itself is subtracted before registration. One should also be aware of the large amounts of data generated with this type of system, and make sure to systemize and automate follow-up and generation of statistics based on the collected data.
- Weighing waste after collection.
  Another strategy is to collect weights of waste after collection, for example in the entrances to waste treatment facilities. Modern treatment facilities are commonly equipped with scales in the entrance and exit the amount of waste unloaded from each vehicle can then be calculated as the difference between entrance and exit weights. However, this might not always be feasible if there is a need to follow up generation of different waste fractions in cases where multicompartment vehicles are used.
- Use of default density values.
  Changes in waste generation can also be evaluated using default values for density and calculation of the number of wastebins emptied. However, it should be acknowledged that changes in recycling behavior could change the density in residual waste and that mere focus on the number of emptied bins might give a distorted picture of the actual situation.
- Self-reported recycling rates
  Several studies have been performed using questionnaires to assess household waste recycling behavior. This method requires direct participation of the households, which might be difficult to achieve. Thus, a low answering

frequency in such studies is not uncommon, and it is, therefore, difficult to assure that the answers obtained are representative of the total population (Mee et al. 2004; Timlett and Williams 2009). Previous studies have shown that responses to questionnaires are commonly subject to the respondent's perceptions of social desirability (Michelson 1990). This means that the person answering the questions often replies according to what he/she assumes to be the correct answer according to social norms or values, rather than his/her actual behavior. The respondent's perceptions of social desirability can also result in a higher answering rate amongst households that are actually participating in waste recycling—assuming that this is seen as socially desirable—while non-participating households choose not to answer the question. In either case, the picture gained from the questionnaire will differ from the actual situation.

• In-depth Interviews with households
Interviews with households regarding their waste sorting behavior can reveal the reasons behind figures generated through technical evaluation methods. The method opens up the possibility of follow-up questions, and decreases the risk of blanks or contradictory replies, which can be the result obtained using questionnaires. However, the method is very time-consuming, as at least 30 min must be allocated to each interview. The number of respondents consulted that can be used to represent the larger population is, therefore, commonly low. According to Esaiasson et al. (2007), at least 10 respondents from each group must be consulted.

• Use of waste composition analyses.
Waste composition analyses can give a detailed view of household waste recycling behavior. Different from any of the above methods, waste composition analyses can give a view not only the amount of waste sorted into different fractions, but also the amount of potentially recyclable material left as residual waste, as well as the quality of the materials actually on-site separated. Based on waste composition analyses, several key data can be generated:

– Specific waste generation (kg household^{-1} year^{-1}).
– On-site sorting ratio (weight %) defined as the weight of collected on-site-sorted recyclable material in relation to the sum of the same material—sorted, incorrectly sorted, and unsorted.
– Ratio of mis-sorted waste (weight %), defined as the mass of nonpackaging, non-newspaper, and nonfood waste deposited in bins designated for the same fractions (Fig. 4.13).

In order to provide information for consistent evaluation and follow developments over time, it is essential that the same method for waste composition analyses is used on each occasion. The method should be consistent both in the way material for analysis is selected, prepared and stored, the way any compensating factors are used and in relation to the choice of fractions and subfractions used in the analysis. Table 4.2 provides an example of the analytical scheme used in waste composition analyses performed in Malmö. The fractions were selected based on the interest in determining the on-site separation of packaging, food

Fig. 4.13 Graphical illustration of the terms on-site sorting ratio (*left*) and ratio of mis-sorted waste (*right*)

waste, hazardous waste, and e-waste. Note that the data provided in the "Additional information" section is essential for accurate calculation of specific waste generation, on-site sorting ratio, and ratio of mis-sorted waste.

However, waste composition analyses are also connected to potential drawbacks. In performing waste composition analyses, there is also a risk of making several types of sampling errors. Pierre Gy in Pitard (1993) list seven different types of such errors; *long-range heterogeneity fluctuation error* (representativeness of a sample for a larger area), *periodic heterogeneity fluctuation error* (not considering potential seasonal fluctuation), *fundamental error* (the size of the sample is too small to account for differences in physical character in the analyzed material), *grouping and segregation separation error* (the sample can be misleading due to uneven distribution of variation in the material), *increment delimitation error* (incorrect division of material into subsamples other than those where the mass center is allocated), *increment extraction error* (material from a sample is lost during subsampling and analyzing) and *preparation error* (changes occurring during the analytical period affects the sample) (Pitard 1993). Several of these errors are related to the fact that the method is work-intensive and thereby costly. It is, therefore, tempting to minimize sampling sizes and assume that samples from a smaller population are representative of a larger population.

For sample sizes when performing waste composition analyses, both absolute and relative criteria have been presented. The amount of analyzed material should be in the range of 450–950 kg according to Petersen (2004), and above 5 % of the population or 100–200 households according to Nordtest (1995) for assessment of specific waste generation and composition analysis respectively. However, some fractions are more prone to fundamental errors than others. If the number of objects belonging to a specific waste category is small, the relative impact on the estimation of occurrence of waste belonging to a specific category and the risk of

Table 4.2 Example of analytical scheme used in waste composition analyses performed in Malmö

Main fraction	Subfraction	Weight (kg)	Percent (%)	kg/household/week
1 Biowaste	Food waste			
	Garden waste			
2 Packaging	Newspaper			
	Paper packaging			
	Plastic packaging			
	Glass packaging			
	Metal packaging			
3 Nonpackaging material[a]	Paper			
	Plastic			
	Glass			
	Metal			
4 Hazardous waste				
5 E-waste				
6 Other waste				
Total packaging waste				
Total waste				
Additional information:				
1 Number of wastebins/amount of waste analyzed[b]				
2 Total number of wastebins/amount of waste of this fraction				
3 Total number of households connected to analyzed waste				
4 Number of days after last collection				
5 Date of analysis				

[a] Nonpackaging waste of made from plastic, metal, paper/cardboard or glass, i.e., the same materials as the ones included in the producer responsibility legislation for packaging materials. Some examples are plastic toothbrushes, metal pots and pans as well as books
[b] In many cases, only a subfraction of the total amount of generated waste will be analyzed. This must be considered in the evaluation of the results from the analysis

making fundamental errors increases. Hazardous waste and e-waste can be used as examples as these fractions commonly represent a small fraction of total household waste generation.

In order to reduce the risk of making fundamental errors, one can either increase the sample size or increase the homogeneity of the waste. The latter can be used in a determination of the waste composition on a chemical basis, but it is not recommended for determination of the on-site separation ratio and he missorted material ratio. If such information is wanted, increased sample sizes are a more viable option. Thus, in order to determine the total amount and on-site separation ratio of less abundant waste fractions, such as e-waste and hazardous waste in solid household waste, the recommendations presented by Petersen (2004) and Nordtest (1995) might not be valid. There is also a current lack of international agreement on standards for methods to perform the analyses. According to Dahlén (2008), data from waste composition analyses performed in different areas are usually impossible to compare.

4.3.2 Caution in Relation to Evaluation of Separate Collection of Food Waste

Food waste is different from many other waste fractions found in solid household waste due to the high moisture content. This can cause moisture losses, both prior to and after disposal of household food waste. This can be particularly relevant significant in cases where household recycling behavior related to food waste is to be investigated. Different types of collection systems for household food waste can of course result in different risks of moisture loss in the collection/transportation phase. As an example, the most common system for separate food waste collection in Sweden is based on the use of paper bags. When food waste is disposed of in paper bags, humidity will naturally evaporate and the weight of the waste will decrease. If the weight loss from separately collected food waste is large—while no or little weight is lost from food waste collected as residual weight (i.e., not on-site separated by household), adjustments are needed in order not to:

* Underestimate the amount of total food waste generated by households.
* Underestimate the household on-site separation ratio (see definition above).
* Overestimate the ratio of mis-sorted material (see definition above).

Previous studies have estimated the weight reduction in household food waste separately collected in paper bags at 19 %, while the reduction was only 5 % when food waste was disposed of in plastic bags (which commonly is the case for non-on-site separated food waste) over a period of 7 days (Bernstad et al. 2012a). In a situation where the on-site separation ratio has been estimated at 50 % and the ratio of mis-sorting at 5 %, these differences in weight loss would mean an underestimation of on-site separation of 6 and 19 % overestimation of mis-sorting. This should also be taken into consideration when different types of systems for separate collection of food waste are compared—such as collection in paper bags versus collection in plastic or bio-plastic bags. Defining on-site separation ratio and mis-sorting in food waste disposal systems is challenging. It is in some cases difficult to compare the influence of different types of technical systems on food waste separation behavior.

4.3.3 Life-Cycle Assessment as an Evaluation Tool

Modern waste management systems are often characterized by processes resulting in both contribution to and avoidance of negative environmental impacts. Energy and resources are often needed initially in the management chain, while avoidance is often achieved in later parts of the treatment system. Thus, it can be difficult to have a clear view of net environmental impacts from different waste management treatment alternatives. Life-cycle assessment (LCA) can be used as a method for comparison of different alternatives for solid waste management (Kirkeby et al. 2006). Thus, LCAs can be used as decision-support tools both for national as well as local decision-makers in the solid waste management area.

The LCA methodology aims to include all relevant significant environmental impacts of the system assessed. However, in practice, an LCA always requires a drastic simplification of a complex reality, where some processes are necessarily left unconsidered. However, it is important to remember that an LCA should include both direct and indirect impacts (modified after Gentil et al. 2009):

- Direct emissions, directly linked to the waste management, produced by collection/transportation, treatment and post-treatment of the waste.
- Indirect emissions or avoided emissions occurring outside the actual treatment system:

  – Upstream activities, for example production of materials and energy bearer or construction of treatment facilities used in the treatment chain.
  – Downstream activities, for example avoided emissions when substituting materials and energy bearers by activities in the waste management chain.

The LCA method is commonly connected to large amounts of data, which is why a number of different computer programs for assessment have been developed over the years: the IWM-model (White et al. 1995; McDougall et al. 2001), ORWARE (Dalemo et al. 1997) and EASEWASTE (Kirkeby et al. 2006), to mention a few. Several documents have previously also been presented to guide users of LCA methodology for evaluations within the waste management area. Some of them are related to specific waste fractions, while others consider solid waste management in general (Table 4.3).

Energy is used in several different parts of the solid waste management chain, both in collection/transportation processes and in treatment processes. Waste management can also result in energy production (for example through incineration), which could substitute other energy bearers. Energy used and substituted in the waste treatment system is relevant significant both in quantitative and qualitative terms, i.e., both the amount and type of energy used is of relevance. Two main approaches can be taken:

- With an attributional approach, energy bearers used in and substituted by the treatment are generally represented by a national or regional average.
- With a consequential approach, the LCA practitioner commonly uses marginal data. With a short term perspective, it could be assumed that the marginal technology is the most polluting and most expensive.

There exists no common view on which approach is preferable in assessments of waste management systems, and the two approaches are currently used in parallel by different LCA practitioners (Mathiesen et al. 2009).

In performing LCAs, the user must choose which environmental impact categories are to be taken into consideration. The relevant significant categories should be considered in each specific case, and, in many cases, LCAs are reduced to taking into account only a few impact categories. According to the European life-cycle incentive (ILCD), the following categories are generally used in LCAs of solid waste management systems (JRC 2011a):

Table 4.3 Examples of existing guidelines related to LCA of solid waste management

Reference	Waste types in focus	Included treatment technologies	Comment
Bjarnadottir et al. (2002)	Biowaste	Incineration, landfills, composting, AD, bio-cells	Includes generic LCI data on emissions from landfills, composting, AD and incineration
JRC (2011a)	MSW	Incineration, landfills, composting, AD, MBT	Pyrolysis and gasification are described without details
JRC (2011b)	Biowaste	Incineration, landfill, composting, AD, MBT	Gives general recommendations on general simplification rules
JRC (2011c)	Construction & Demolition (C&D) waste	Not applicable	Includes composition of C&D and outline of key principles in performing LCA of C&D waste management
Sundqvist (1999)	MSW, industrial residues (coal ashes)	Incineration, landfills, bio-fills, cell deposits	Gives recommendations on sound allocation and system boundary setting methodologies.
la Cour Jansen et al. (2007)	Biowaste	MBT, incineration, composting (windrow, closed reactor, home composting), AD, landfills	Includes LCI data on landfills, composting, AD, incineration, MBT
PCR (2008)	MSW	Incineration, landfills, composting, MBT	Includes LCI data on energy content, biogas production and collection

- Climate change;
- Ozone depletion;
- Human toxicity;
- Particulate matter/Respiratory inorganics;
- Ionizing radiation;
- Photochemical ozone formation;
- Acidification;
- Eutrophication;
- Ecotoxicity;
- Land use;
- Resource depletion;
- Noise, odor, accidents, desiccation, erosion, salination, soil quality, etc.

However, reviews of a large number of precious LCA studies of solid waste management systems show that less than 30 % include both the nontoxic and toxic impact categories suggested by the ILCD (JRC 2011a), while they all considered the category, global warming (Bakas et al. 2012).

A general difficulty in use of LCA is the problem of addressing environmental impacts that are difficult to translate into numeric values. An example is the difficulty of addressing resource depletion—such as use of nonrenewable energy or minerals or long-term degradation of farmland. Also noise, odor, working conditions, and soil quality are difficult to address. Attempts to address these issues have been made through monetized environmental burdens in cost-benefit analyses. However, such evaluation-systems have also been criticized due to a limited scientific basis for such monetization of environmental impacts (Stirling 1997).

Thus, an LCA is always connected to a large number of uncertainties, and it is important for both users of the tool and those who use the results to know that the information it generates is neither complete nor absolutely objective. It must, therefore, be kept in mind that LCAs can be interesting for evaluation of different solutions for solid waste management, but as a decision-support rather than a decision-making tool. Other concerns, such as economic resources, ethical issues, and social willingness must indeed also be considered when making decisions related to solid waste management (Kirkeby et al. 2006).

4.4 Project and Collaboration Structure

It has previously been stated that there is often a gap between the academic research connected to management of solid household waste and the local governments and authorities where the management system is being planned (Parfitt and Flowerdew 1997). Malmö has for several years searched to bridge this gap through close collaboration with academia. In the following, the collaboration structures of the different projects discussed in this text are described.

4.4.1 The Bo01 Project Organizational Structure

The decision regarding waste management systems introduced in the area was, to a great extent, the outcome from a process where several other actors involved in the housing fair were involved; the Department of Parks and Public Spaces, the Department for Environmental Protection, the Office for Urban Planning as well as the steering group for the housing fair. Solid waste management was, during the planning phase of the Bo01 project, managed through a waste management department under the Department of Parks and Public Spaces and, the initial ideas from the waste department were, to a great extent, never realized. The evaluation of the systems in the Bo01 area was made through several parallel projects, co-ordinated by the Department for Waste Management, which, at this point, had been merged with the Department for Water Management. Evaluations were made in close collaboration with the regional and municipally owned treatment enterprise, Sysav as well as with academia. These projects were related both to household behavior, led by the University of Gothenburg, as well as to technological functions led by Lund University.

4.4.2 The Augustenborg Project Organizational Structure

A project group was formed consisting of VA SYD, MKB and Lund University. Also the waste collector entrepreneur participated in the project group, albeit not on a regular basis. Meetings with all participating agents in the project group were held on a regular basis around every sixth week. Participants in the groups were public servants directly involved in activities performed within the project. On the university side, a Ph.D. student participated in the project group. The steering committee consisted of executives from VA SYD, MKB, Sysav and Lund University. Members of the steering committee could, thanks to their broad experience, contribute with a different view on the difficulties encountered in the project. The steering committee met four times per year. In order to bring information from the project group level to the steering committee, representatives from Lund University participated in both groups.

The structure created a platform for direct communication among all agents connected to different parts of the waste management chain. Problems, and the often differing interests of the agents, could be raised and discussed in order to find sustainable solutions, where solving one problem does not lead to a new one in a different part of the chain. Thus, in correlation to the life-cycle analysis approach, a holistic approach was also sought at the organizational level of the project. In order to increase the interaction with external actors, a reference group was created, consisting of the Swedish Waste Management Association, the Swedish organization for municipal facility owners, SABO, amongst others. Also, further collaboration with the university was gained through the inclusion of the Institution for Ethnology, which performed far-reaching interviews in the area during the project period (Table 4.4).

Table 4.4 Organizational structure of the Augustenborg project

Project group	Steering committee	Reference group
VA SYD	VA SYD	Swedish Association of Public Housing Companies
MKB	MKB	Swedish Waste Management Association
Lund University	Lund University	Environmental Department, City of Malmö
Collection entrepreneur	Sysav	Larger facility owners
		The Swedish Tenants Association (SABO)
		Representatives from households in the area

4.4.3 The Role of Academia

In the case of the Bo01 project, academia was given the role of external evaluation of the systems for waste management introduced in the area, providing independent and unbiased evaluations of the systems and interferences in the area.

In the case of the Augustenborg project, academia had an even more prominent role. The organizational structure provided close contact among the different actors involved in the waste management chain and academia. This is important for several different reasons:

- For the success of the projects as such, continuous structured evaluation of different changes made in the system can give direct feedback to the waste management organizations, which can rapidly plan their activities in relation to given signals.
- For the quality of the academic work. This can be improved when academia has great knowledge of the practical arrangements of the waste management system, direct contacts with the actors involved in the process, as well as full insight into any factors which might be influencing results or of relevance to the evaluation of these.
- For the understanding of results provided by academia within the waste management organizations. This can be improved when persons within these organizations are well informed about how the results have been obtained. Methods and data used in evaluations can be discussed among the different actors involved in the project so that all parties involved understand how this can affect the results obtained. This can be important when outcomes are used as decision support and can make nonacademic parties feel more secure in using and presenting results externally as well as internally in their organizations.

4.5 Project Outcomes and Learning

4.5.1 The Bo01 Project

The waste management system introduced in the Bo01 area was evaluated in several phases. A general difficulty in the evaluation of the project was that it took rather long before the apartments in the area were rented. Thus, the number of inhabitants in the area was still low in the first evaluation phases of the project (2000–2003). The second evaluation phase (2009–2013) only included the technical performance and user-friendliness of FWD.

4.5.1.1 On-Site Separation of Food Waste with Unconventional Food Waste Disposal Systems

Results from analyses of collected sludge from the tank systems showed that the sludge TS was very low, in all cases less than 3 %, and in the system with a cutting pump (area A), less than 1 %. (Gruvenberg et al. 2003). This made the transportation of sludge to treatment plants rather inefficient. The concentration of fat in collected food waste sludge was high, showing that the risk of fat related problems in the sewage system was vastly decreased through the use of settling tanks. The high amount of fat in collected food waste sludge also resulted in high production of biogas in the methanogenic batch tests performed. However, it was also seen that the effluent contained rather high concentrations of organic matter, nitrogen, and phosphorus. It was suggested that hydrolysis takes place in the tanks, resulting in partial degradation of the food waste prior to collection (Bernstad et al. 2012b). Thus, the system demonstrated several benefits. However, an optimization of the system could increase benefits even further, and several suggestions were made for slight changes in the design of the system prior to new installations in the city. These included, among others, use of FWD where the particle size of ground food waste was increased. Continuous collaboration between academia and the Department for Waste Management will secure systematic evaluations of these new systems in order to investigate how these changes have affected the performance of the system.

4.5.1.2 On-Site Separation of Food Waste with the Vacuum System

A total of five waste composition analyses were performed in order to evaluate the on-site separation ratio and mis-sorted materials ratio in waste disposed of in the vacuum system. Mis-sorting ratios were initially high—varying between 12 and 16 %. Due to this, separately collected food waste from the area was re-classified as residual waste for incineration instead of anaerobic digestion. After the fourth analysis, a leaflet was sent to all households, with a clear message about the necessity to keep this fraction clean. A decreased ratio of mis-sorting in the fifth composition analysis could be a result of this campaign. It could also be partly explained by less engaged households no longer participating in the on-site separation, as the last analysis also showed a decreased on-site separation ratio amongst the households. Two years after installation, less than 30 % of the organic waste was separately collected with this system.

Visual inspections of food waste separately collected through the vacuum system made at the receiving treatment plant showed that the quality remained low. Thus, in 2009, locks were installed on all inlet doors to food waste chutes. This improved the quality of the separately collected food waste, and, since then, it has been possible to use food waste collected in the area for biogas production. However, this change has had negative effects in relation to user-friendliness, as seen below.

4.5.1.3 User-Friendliness of Novel Waste Management Systems

Evaluations of the systems from the perspective of the households have been made in two rounds; once soon after the systems were installed (in 2002–2003) (Åberg 2004) and once in 2011 (Hoppe 2013).

The evaluation performed in 2002–2003 consisted of two rounds of in-depth interviews with households in the area. Prior to the interviews, 43 households were contacted through a letter presenting the aim of the study. A week later, the same households were contacted through a phone call, in which they were asked whether they wanted to participate in the study or not. 30 of them were willing to participate.

Households connected to the food waste disposal system were, according to the interviews performed, in general, satisfied with the system. Households reported that the FWD facilitated the kitchen work and the system was, in general, seen as easy to handle. However, in some cases, households reported that the sound from the grinders were too loud, that bad smells could escape from the grinder tubes, and that on/off buttons, in some cases, were difficult to reach at the same time as disposing of food waste in the sink.

Households connected to the vacuum system were, to a greater extent, of the opinion that the separate collection of food waste was messy, and they were troubled by the functionality of paper bags and afraid that these would not cope with the weight of the heavy wet food waste. Households also had trouble finding new paper bags in the areas where these should be found (Åberg 2004).

In 2011, the two systems for food waste collection introduced in the Bo01 area were again evaluated from a user perspective through semi-structured interviews (Hoppe 2013). However, this time, households in a more recently constructed high-rise building in the area, where FWD had been installed, were included in the study.

Interviews with households using FWD show that tenants often choose not to use the grinder due to the sound it makes, especially at specific times of the day (i.e., early mornings and late evenings) and especially not for particular types of food (i.e., hard/stiff types of food waste, such as bones, carrots, etc.). This reduced the use of the FWD rather drastically in some households. The interviews also show that tenants in many cases had little knowledge of the system—both of what can and cannot be disposed of in the FWD as well as of what happens after grinding—even when having been provided with information materials about this (brochures).

In relation to the vacuum system, interviews show that, in some cases, households were not participating in the food waste separation due to very banal reasons, such as having only one key to the chutes, which the households chose to leave in the house. Also, according to the interviews, fear of smell and leakage, as well as having to take the paper bags out extra space were seen as disadvantages of the vacuum system, while FWD were viewed as ergonomically inferior when disposing of larger amounts of organic waste at the same time, some specific types of food waste and at specific times of the day. Thus, although FWD in many cases have been seen as very convenient, they can be associated to several less beneficial aspects and are not a guarantee of high participation rates and on-site separation ratios (Hoppe 2013).

Thus, the two evaluations show that problems related to use of both food waste grinders and vacuum systems, from a user perspective, are rather similar. Problems related to paper bag collection remains, and outcomes from the first evaluation round have not resulted in learning, demonstrated in concrete improvements in the second installation of the food waste disposal systems in the high-rise building.

4.5.1.4 Organization and Maintenance

The vacuum system installed in the area connected several different property owners to the same infrastructure for collection of household waste. Thus, ownership and responsibilities in relation to maintenance had to be organized and defined. An association was created by all property owners with buildings that were to be connected to the vacuum system. Investment costs were shared among the different members of the association. The tanks in which bags for separately collected food waste and residual waste are collected prior to transportation in vacuum pipes to collection points are owned by each facility owner, while the pipes to the collection point are owned by the municipality. Ten years after the installation, the pipes were in need of refurbishment. This was financed by the municipality. Also, the tanks have been in need of refurbishment, which can become rather costly for facility owners, and a change in the collection of food waste in separate food wastebins have, therefore, been discussed as well as long-term service contracts with the supplier of the vacuum systems, which could reduce the costs for facility owners.

An unexpected consequence of the vacuum system was that large parts of what normally was considered "common" residual waste could no longer be disposed of as such, as it would not fit into the chutes in the vacuum system, and so it had to be disposed of as bulky waste. With the closest recycling station for bulky waste at a distance of several kilometers from the Bo01 area, this highly increased the risk of waste dumping in the area.

This process clearly shows the need for transparency of the life-time costs of different systems—including necessary maintenance, refurbishments and increased costs due to waste dumping. It also shows that it might be wise to have a single owner of a system or arrange a system for commonly financed maintenance of the whole system, if the ownership is divided.

Another issue related to the organization of solid waste management in the Bo01 area is related to the liberalization of collection of waste under the producer responsibility legislation. This means that the collection service for such waste is to be contracted on the free market. Initially, the contracting of collection services was co-ordinated by the municipality. However, after the finalization of the first contract, each facility owner was free to choose his own enterprise for collection of dry recyclables. This created a great risk of counteracting the objective to minimize the circulation of heavy vehicles in the area and decrease the need for waste transportation.

4.5.1.5 Information Provided to Households

The information campaigns performed among households connected to the novel systems for food waste collection when the systems were installed, were evaluated through in-depth interviews with a total of 30 households in the area in 2002–2003 (Åberg 2004).

The majority of the households recalled seeing the brochure that had been sent to all households with information about the food waste collection systems in the area. However, the degree to which the households had taken part in the information is unclear, as many of the households interviewed asked for more information that had, in fact, been given in the brochures. The information missing from the perspective of the households was, in many cases, related to what happens to the food waste after collection. While most households were positive toward the oral information provided through the doorstep campaign in the area, some of the households regarded this as "overdoing", as households had felt patronized and viewed it as an unnecessary expense on the part of the municipality. None of the households had taken part in the information about food waste management in the area through the municipality's information sites on the Internet. Several of the households reported that they had looked for and found information from facility managers and facility owners (Table 4.5).

Table 4.5 Information channels through which households participating in the interviews report to have obtained information about food waste management

Information channel	Percent
Brochure	88
Oral information	58
Internet	42
Other[a]	17

[a] Facility managers and facility owners

The "information acquired through others" is interesting and gives important indications that it could be important to inform also these groups about the systems, so that they can pass correct information on to households. In relation to this, the researcher responsible for the evaluation also observed that information on the recycling buildings in the area were provided from three different senders; the Municipal Department for Environmental Protection, the Municipal Department for Waste Management and the entrepreneurs hired for the collection service, and that the information from the different senders, in some cases, was conflicting.

From a comparative point of view, the evaluation stated that households connected to FWD were, to a lesser extent, aware of what happened to the food waste after disposal compared to households connected to the vacuum system. However, knowledge was low in both cases (Åberg 2004).

4.5.2 The Augustenborg Project

4.5.2.1 On-Site Separation of Dry Recyclables

Waste composition analyses performed in the Augustenborg area were used to assess the overall composition of the waste, as well as recycling behavior (Table 4.6).

As seen in Table 4.6, the highest on-site separation ratios are reached for the fractions, glass and newspaper. These were also the fractions where the producer responsibility legislation system was introduced and where Swedes have the longest tradition in relation to waste separation and recycling. These are also the fractions where the lowest mis-sorting ratio is seen.

As the Swedish system for material recycling is based on producer responsibility legislation on packaging, recyclable products, such as plastic toothbrushes and metal pots and pans are currently considered as being nonrecyclable. Actually, in the waste composition analysis method presented previously, a plastic toothbrush among plastic packaging would be considered as mis-sorting. This might not always be clear and logical, neither from the perspective of households nor from an environmental point of view. Thus, a change from the current sorting system, based on the producer responsibility legislation on packaging material to a system

Table 4.6 On-site separation ratio and ratio of incorrectly sorted material in dry recyclables (weight %) in the Augustenborg area, as averages based on three waste composition analyses

Waste fraction	On-site separation ratio (%)	Ratio of mis-sorted material (%)
Clear glass packaging	73	5
Colored glass packaging	83	11
Metal packaging	38	30
Plastic packaging	40	32
Paper packaging	55	29
Newspaper	70	5

based on sorting in material streams has been discussed and suggested by the Swedish EPA (SEPA 2007).

The waste composition analysis method used in the Augustenborg project provides an opportunity for investigation of potential impacts due to a change from packaging recycling to material stream recycling. Based on results from this case study, a change would increase potential recycling from 80 % of total waste generation with the current packaging recycling system to 92 % with a material stream recycling system. Results also show that a change would have a great impact in relation to metal, while the impact in relation to cardboard is small, as the amount of nonpackaging metal waste is high in relation to the amount of nonpackaging cardboard waste (Bernstad et al. 2011).

Results from waste composition analyses also show a correlation between low on-site separation ratios and high ratios of mis-sorting among dry recyclables. This was interpreted as a greater uncertainty in relation to some fractions (i.e., plastic and metal packaging) compared to others (newspapers and glass packaging). According to results gained from the ethnological studies performed in the Augustenborg area, households are inclined to do nothing rather than do wrong; i.e., if households are uncertain of how to sort a specific packaging, it might be easier not to sort it at all (Ewert et al. 2009).

4.5.2.2 Environmental Impacts from On-Site Separation of Dry Recyclables

In order to evaluate the environmental effects of the on-site recycling scheme for dry recyclables in the Augustenborg residential area, a life cycle assessment was performed. The overall questions in the assessment were related to the environmental benefits from the current system and the potential benefits with an optimization of the recycling behavior (Fig. 4.14). Data from waste composition analyses performed in the area were used as input data in the assessment. Thus, site-specific data was used both in relation to waste composition and on-site separation ratios for different types of dry recyclables. In order to assess environmental impacts from recycling and treatment processes, it was assumed that all materials were recycled in Sweden. Thus, transportation distances to bigger

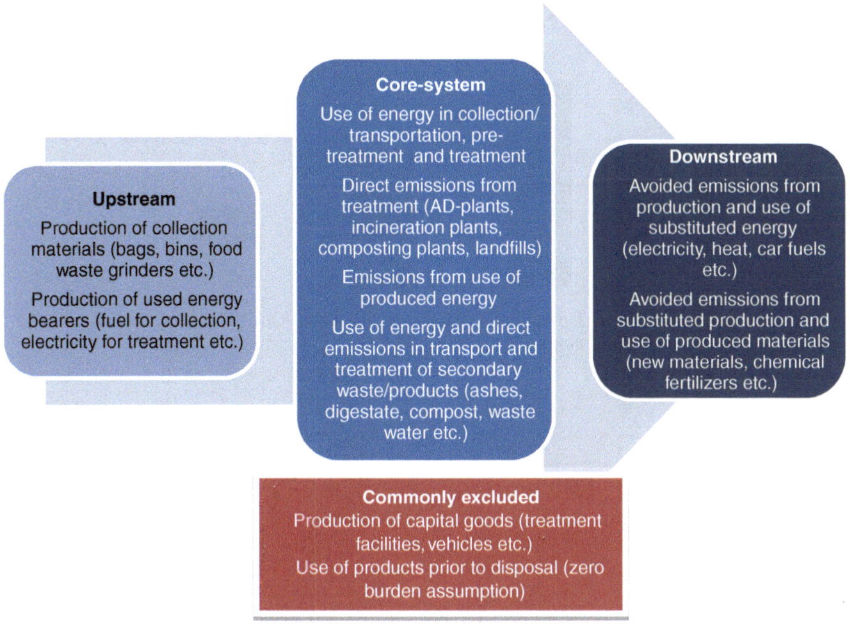

Upstream

Production of collection materials (bags, bins, food waste grinders etc.)

Production of used energy bearers (fuel for collection, electricity for treatment etc.)

Core-system

Use of energy in collection/ transportation, pre-treatment and treatment

Direct emissions from treatment (AD-plants, incineration plants, composting plants, landfills)

Emissions from use of produced energy

Use of energy and direct emissions in transport and treatment of secondary waste/products (ashes, digestate, compost, waste water etc.)

Downstream

Avoided emissions from production and use of substituted energy (electricity, heat, car fuels etc.)

Avoided emissions from substituted production and use of produced materials (new materials, chemical fertilizers etc.)

Commonly excluded

Production of capital goods (treatment facilities, vehicles etc.)

Use of products prior to disposal (zero burden assumption)

Fig. 4.14 Graphical representation of direct and indirect processes considered in an LCA of solid waste management

recycling plants, as well as emission data from Swedish plants on material recycling and treatment, were used (Bernstad et al. 2011).

An attributional approach was used in the study. Thus, energy bearers used as well as produced in the treatment, were represented by national average electricity generation and regional average heat generation. From a Swedish perspective, the electricity is, to a great extent, provided through hydropower and nuclear power. The regional heat is provided mainly through waste incineration, surplus energy from industries and bioenergy. This means that used and substituted electricity and heat make a very low contribution to global warming. However, with a consequential approach, the marginal electricity and heat production must be identified. According to Mathiesen et al. (2009), it can be assumed that the marginal energy production is equal to the most polluting and most expensive energy production. Due to the Swedish carbon dioxide tax, it can be assumed that marginal energy production is related to use of fuels with high impacts on global warming. Thus, with a consequential approach and a short-term perspective, both used and substituted energy can be assumed to be represented by the use of coal in combined heat and power plants (Elforsk 2008).

Results show that the current waste management system results in net avoidance of global warming potential from the waste management system applied in the Augustenborg residential area. However, the potential to increase the benefits are vast (Fig. 4.15). The main reason for this is that the fractions with the lowest on-site separation ratios also are the fractions related to the lowest environmental

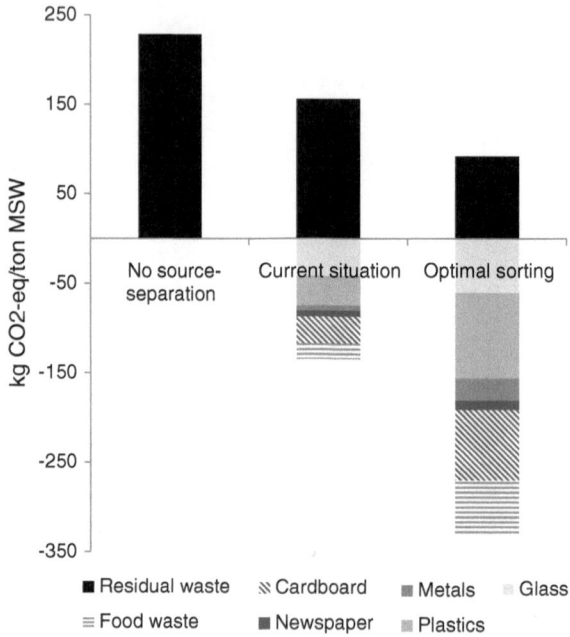

Fig. 4.15 Net GWP from the waste management system applied in the Augustenborg area without any material recycling, under current conditions and under optimal conditions (kg CO_2-eq./ton generated waste)

benefits. Sensitivity analyses were performed in order to assess the impact from the assumptions made in relation to environmental impacts from energy used and substituted in the waste management system. With a consequential approach, benefits from incineration are increased due to high efficiency in energy recovery in modern Swedish waste incineration plants and a well-developed system for district heating (Fig. 4.15).

Also, material recycling processes are affected by change from the attributional to the consequential approach. As an example, energy used for paper recycling is, to a great extent, in the form of electricity, while energy consumed in the production of new paper is commonly derived from the combustion of residuals from the forestry—i.e., bioenergy. Thus, with a consequential approach and looking only at the impact category, global warming potential, it can be difficult to defend paper recycling. However, it could be stated that bioenergy used in the virgin new paper production could be used to substitute fossil energy bearers in the energy system, which again could change the balance of the system. It is important to point out that paper is an exception in this case. In the case of plastics, metals and glass, the benefits in relation to global warming are substantial, independently of the assumed environmental profile of energy used in recycling processes.

Potential environmental benefits from the change from packaging recycling to recycling of material streams were investigated using LCA. In the assessment, it was assumed that environmental impacts and benefits related to recycling of nonpackaging materials were equal to the ones related to recycling of packaging materials. Results show that relative environmental benefits related to material stream recycling are higher than the relative decrease in residual waste generation this would lead to (i.e., a potential 12 % decrease in residual waste generation). The reason for this is the higher environmental benefits per ton of recycled material in the case of plastics and metals—which account for the largest amount of nonpackaging recyclable waste, compared to cardboard and glass, which correspond to a smaller fraction of the total material stream of separated waste (Bernstad et al. 2011).

4.5.2.3 On-Site Separation of Hazardous Waste and Electronic Waste

The introduction of a system for on-site separation of hazardous waste and electronic waste (e-waste) resulted in a clear reduction of hazardous waste and e-waste disposed of as residual waste and waste fractions destined for packaging recycling in the Augustenborg area. The amount of incorrectly disposed hazardous waste was reduced from 1.2 to 0.6 kg per household per year, while a reduction from 2.5 to 1.2 kg per household per year was seen in the case of e-waste. At the same time, 1.3 kg hazardous waste and 8.8 kg e-waste per household per year were separately collected. The facility managers experienced fewer problems with dumping of bulky e-waste in basements and other places in the residential area, and the workload of facility managers was reduced.

Results imply that a successful management system for on-site separation of larger e-waste items might not be optimal for on-site separation of smaller devices, as very few smaller and more valuable e-waste items were deposited in the designated cages. Thus, a further development of cage design, with a smaller box for small e-waste was developed. In performed waste composition analyses, it was seen that a considerable amount of non-on-site separated e-waste was incorrectly sorted as metal or plastic packaging. As an example, electronic toys made from plastic were commonly found in bins for plastic packaging, while lamps, etc., made from metal were found among metal packaging. This indicates that households tend to sort e-waste according to the material used in the visual outer parts of the item, rather than the fact that it contains electrical components containing both precious and potentially hazardous compounds. Thus, additional information to households may be needed in order to improve the on-site separation even further (Bernstad et al. 2010).

4.5.2.4 On-Site Separation of Food Waste

In order to evaluate the potential environmental benefits from introduced changes in relation to the food waste management in the Augustenborg area, incineration, decentralized composting and centralized anaerobic digestion of household food waste are compared using lifecycle assessment methodology. To the largest possible extent, the modeling was performed using site-specific data for transportation distances, vehicle emissions, potential energy recovery, input energy use, treatment emissions and potential nutrient recovery from the Augustenborg residential area.

Results show that both biological treatment methods assessed in the study— centralized anaerobic as well as decentralized aerobic, result in net avoidance of greenhouse gas emissions, but make a larger contribution both to nutrient enrichment and acidification when compared to incineration. Use of biogas as a substitute for fossil car fuel results in greater avoidance of greenhouse gas emissions compared to when using biogas fuels electricity generation (Fig. 4.16).

In the study presented above, an attributional approach was used. Applying a consequential approach to the Augustenborg waste management system can quite drastically change the results in relation to the potential for global warming. As seen in Fig. 4.17, incineration becomes much more beneficial using a consequential approach. The main reason for this is the potentially high energy recovery from incineration and assumed high avoidance of CO_2 emissions when substituting Danish coal power, which was assumed to be the marginal electricity production in this study due to the interconnection of the Nordic electricity grids. However, anaerobic digestion is still the most beneficial alternative in relation to global warming, although use of produced biogas for energy generation rather than as vehicle fuel, is assumed to be preferable.

Fig. 4.16 Environmental impacts from anaerobic digestion (using produced biogas for electricity and fuel respectively), incineration and composting of household food waste, displayed as impacts related to eutrophication, acidification and greenhouse gas emissions. Note the interrupted bars in the compost scenario

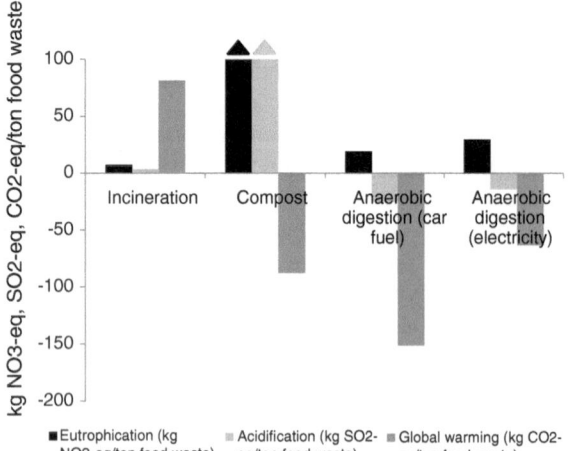

Fig. 4.17 Global warming potential from anaerobic digestion (using produced biogas for electricity or fuel respectively), incineration and composting of household food waste, assuming that consumed and produced energy is represented by average electricity (SE) and marginal electricity (DK)

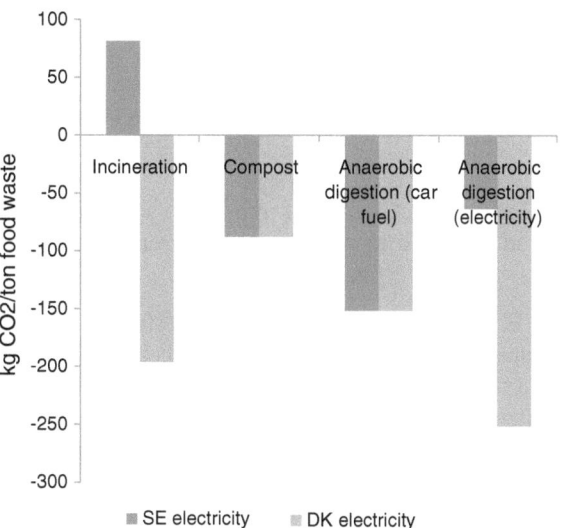

Thus, the outcomes from an LCA study are highly dependent on assumptions made by the LCA practitioner. Due to the uncertainties related to substituted electricity, it could be recommended that produced energy should be used in the transportation sector, where the difference between average and marginal energy bearers is small. In 2010, only 6 % of the total fuel used by cars, lorries and buses in Sweden was based on renewable resources. However, the difference between results obtained when using an attributional and consequential approach clearly shows how important it is to report assumptions made in the assessment.

As previously stated, life cycle assessments can be used not only to compare different treatment alternatives, but also to identify hot spots—i.e., factors resulting in great environmental impacts in the system. Several such factors were identified in the assessment:

- No cleaning of emissions of NH_3, N_2O or CH_4 was conducted in the case of decentralized composts at the study site. Installation of bio-filters could have largely reduced these emissions and thus have decreased greenhouse gas emissions, as well as the contribution to acidification and eutrophication.
- The pre-treatment of separately collected food waste prior to anaerobic digestion was seen to be inefficient in the sense that large amounts of the potentially biodegradable material was separated from the fraction going to anaerobic digestion. Rejected material was instead incinerated together with residual waste. Over 20 % of the potential methane production was lost through the pre-treatment (Bernstad et al. 2013a).
- Use of digestate from anaerobic treatment as a substitute for chemical fertilizers is essential for the environmental benefits related to anaerobic digestion.

- Results are, to a high degree, dependent on energy substitution. It was seen that if it is assumed that produced biogas substitutes Danish coal in electric power generation, this is preferable to the use of biogas as car fuel.
- Use of plastic bags for separate collection of food waste would decrease the overall reduction of greenhouse gas emissions, but the impact is small in relation to other parts of the system.
- Environmental impacts related to collection and transportation plays a very inferior role in relation to the overall environmental performance of all the systems investigated.

Based on the above, it is seen that factors controllable by the Municipal Waste Management Department, such as choice of fuel used in collection/transportation and the types of bags used for collection, are seldom decisive for the ranking of different treatment alternatives. On the contrary, major impacts are often dependent on processes allocated far from the control of local decision-makers, for example by farmers' potential willingness to use digestate as a substitute for chemical fertilizers, or car-owners willingness to invest in gas-driven vehicles rather than petrol or diesel-driven ones. This indicates the importance of a holistic approach and extended collaboration between agents in the waste management chain and other sectors of society in order to optimize potential environmental benefits related to waste management.

4.5.2.5 Information for Households

A study was also made on the use of information aimed at households. Written information was delivered to all households when the system for on-site separation of food waste for biogas production was introduced in the area. The printed material contained both information about *how* food waste was to be separated at source in the household (practical information on how to minimize risk of leakage from paper bags or of attracting flies, etc.) and in the recycling building, as well as *why* households were asked to do this (information on environmental benefits from anaerobic treatment of food waste, the amount of produced biogas per kg of food waste, etc.). Also, 66 % of all households in the area received oral information through a doorstep campaign (Group 1) (Fig. 4.18). Informants employed by the municipality provided information focused on the environmental benefits with food waste recycling and how collected food waste is treated after collection. Households were informed that biogas produced from separately collected food waste is used as renewable fuel in vehicles and that remaining biosolids are used as fertilizers on farmland. Informants also delivered a special vessel for food waste separation and a first set of paper bags to all households as a part of the doorstep campaign.

The effects of the campaign were evaluated, both through waste composition analyses (before as well as after the campaign) and weighing of separately

Fig. 4.18 The study site with indication of different information strategies. Group 1. Written and oral information regarding separate collection and recycling of food waste (420 apartments); Group 2. Written information about separated collection and recycling of food waste (210 apartments)

collected food waste over a period of 104 weeks. Also, a questionnaire was sent to all households in the area, 34 weeks after the campaign in Group 1 (Fig. 4.19).

The results based on weekly collection of weights from generated residual waste, on-site separated dry recyclables and on-site separated food waste showed a statistically significant decrease in the generation of residual waste after the scheme for separate collection of food waste was introduced.

Results from the waste composition analyses showed that the average on-site separation ratio of food waste was higher, and the ratio of incorrectly sorted material in the food waste fraction was lower among households where oral information had been provided, compared to households receiving only written information. However, a decrease over time in the on-site separation ratio of food waste amongst households receiving oral information suggests short-lived effects of the door-to-door campaigns. The ratio of impurities in on-site separated food waste was, however, lower amongst households that had received oral information throughout the study period.

The ratio of mis-sorting was initially very high (16.4 %) in Group 2. However, a large reduction in incorrect sorting was seen after a change in the arrangement of wastebins in recycling buildings—resulting in a need for more effort to avoid depositing nonfood waste in bins for food waste and a drastic decrease in incorrect disposal of residual waste among food waste (Fig. 4.20).

According to answers from the questionnaire, 82 % of the responding households reported that they on-site separated their food waste, and 60 % of the households where on-site separation of food waste was self-reported, had received oral information regarding this fraction. Only 17 % of the nonparticipating households that answered the questionnaire had received oral information regarding food waste recycling. Thus, these figures could be interpreted as if the doorstep campaign had had a substantial impact on the on-site separation behavior amongst households. However, only 13 % of all households in the area answered

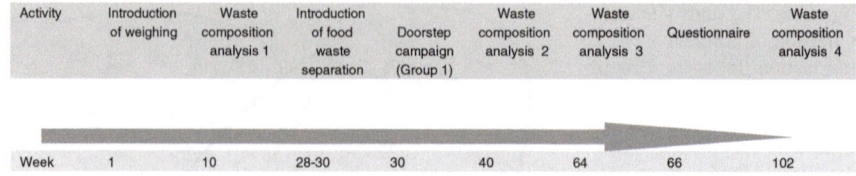

Fig. 4.19 Time line for activities related to waste sorting and evaluation over the study period

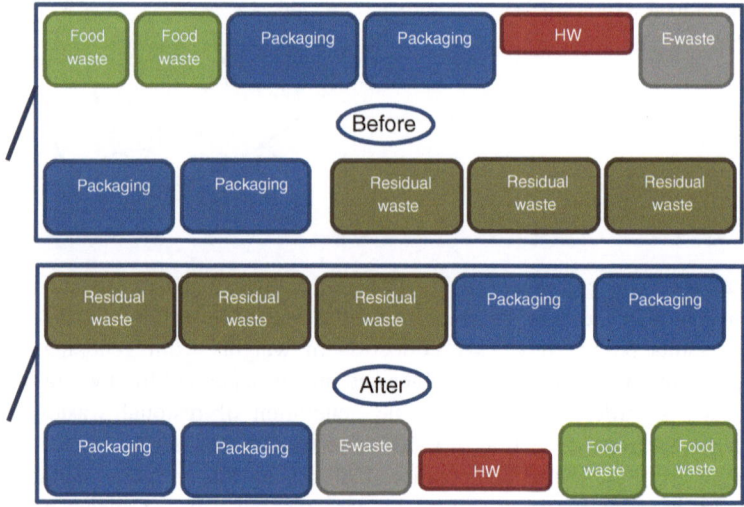

Fig. 4.20 Relocation of wastebins in waste sorting buildings in the Augustenborg area

the questionnaire, and, as the self-reported participation rate was very high compared to the on-site separation ratio reported based on performed waste composition analyses, it is probable that households participating in the food waste recycling were more attracted to answering the questionnaire—irrespective of whether they had received oral information or not. Only around 40 % of the households that had received oral information reported that this had provided them with more profound information about the recycling scheme than they had already received through the written information distributed. This could imply that what is actually said during the oral information might be of less importance than someone demonstrating the importance of the requested behavior. It can also indicate that receiving the material needed to start on-site separation of food waste at home (paper bags and containers) might have been the most important outcome from the oral information campaign, as it was seen that both paper bags and containers left in recycling buildings for households in Group 2, were, in many cases, not collected by the households, even several months after the food waste recycling scheme had been introduced (Bernstad et al. 2013b).

In-depth interviews were also performed with a smaller number of households in the area. Results from the interviews show that, although many households had a positive view on on-site separation of food waste in theory, in many cases, they saw the arrangements as impractical and unhygienic. The vessel for food waste collection was in these cases often placed on the kitchen sink, an area that, in the small apartments in the Augustenborg area, constructed in the early 1950s, is very limited (Ewert et al. 2009). According to the questionnaire, more than 70 % of the households not on-site separating food waste claimed that lack of space in the kitchen were the main reason for this (Bernstad et al. 2013b). Results from the interviews also show that households can regard their on-site separation of waste as a service provided to the facility owner. This "free labor" should, according to them, be rewarded through, for example cuts in rents. If such incentives are lacking, the interest in participation on the part of the households might be limited (Ewert et al. 2009).

Based on this, a project was initialized in order to improve the infrastructure for on-site separation inside the households. The project was developed in collaboration with both Malmö and Lund University. Master students from the Production Design Program at Malmö University were engaged in development of the on-site separation concept so that it could be easily installed in existing kitchens in order to facilitate primarily food waste separation and also separation of recyclable packaging. Several ideas were developed and the final concept, containing a metal hanger for vessels aimed at separate collection of household food waste in paper bags, was produced on a large scale (Fig. 4.21).

A leaflet with information about the environmental benefits related to food waste separation was distributed amongst all households in the area. Also, a doorstep campaign, in which the hanger for food waste separation was installed, and oral as well as written information regarding the environmental benefits from food waste sorting was distributed, was performed amongst 212 apartments in the area.

Fig. 4.21 Metal hanger mounted inside kitchen sink lockers and information material delivered by information staff during the campaign "And around again". *Photo* Gugge Zelander

The campaign led to a sharp increase in the amount of separated food waste during the first 2.5 months after the campaign in households, which had managed to get the hanger for food waste separation installed. After this period, recycling rates declined. The decline coincided with the summer holiday season, and the rates increased again during the fall to levels slightly higher than the ones seen prior to the campaign. However, based on data from the treatment facility, the leaflet sent to all households in the area as a part of the campaign, "An around again" did not result in any general increase in the amount of separately collected food waste in the area.

An evaluation of the campaign was conducted in the fall of 2011 through in-depth interviews with 13 households. The evaluation showed that households were largely very positive about the metal hanger and felt that this made sorting easier. However, the campaign message was unclear to the households and it had been difficult to link the campaign slogan to the environmental benefits from food waste sorting (Malmberg and Nilsson 2011).

In autumn 2012, the metal hanger was installed in all the remaining apartments in the area. All apartments also received bags designed to facilitate recycling of packaging materials. A short and simple slogan was chosen: "Thanks for the food", but the use of written information material was very limited in this campaign. Instead products were supposed to speak for themselves. Waste composition analyses of residual waste and food waste as well as weighing of separately collected food waste were used in order to evaluate the campaign. Waste composition analyses performed before and after the campaign showed that the amount of recyclable materials (i.e., packaging materials and newspapers) in residual waste decreased by 40 %; from an average of 3.1 kg/household per week to 1.9 kg/household per week after the campaign. The effect of the campaign could be assessed on a more long-term basis through weighing of on-site separated food waste. Weighing was performed 2, 4 and 6 months after the campaign. Results display an increase in the amount of separately collected food waste from 1.05 kg/household per week to 1.61 kg/household per week on average. This corresponds to an increase by more than 50 %. The results, therefore, support earlier studies suggesting that accessibility and the level of effort needed in order to participate in recycling are important factors when explaining different levels of participation in waste recycling (Derksen and Gartrell 1993; Martin et al. 2006; McDonald and Ball 1998; Miranda and Aldy 1994), but they also imply that accessibility needs to be studied in a broader sense—also reflecting adequate interior space. A positive relationship between adequate space for recycling inside households and recycling behavior has been stated previously by Ando and Gosselin (2005), but, besides this, it has been poorly investigated and discussed in the academic literature.

Several points learned in the project:

- It is of great importance to provide households with information both on HOW and WHY they should participate in recycling. When introducing new recycling schemes, the HOW information must be detailed to avoid insecurity among households.

- At the same time, providing households with information related to environmental procurement can have little effect on recycling behavior if the information is not combined with physical changes that facilitate the recycling. Thus, providing households with needed physical equipment for separate collection can be highly important when initiating a recycling scheme. This can be combined with doorstep campaigns in order to make sure that the equipment actually is installed in the household.
- The fact that on-site separation starts in the household (kitchen in the case of food waste) must be considered. Thus, the facilitation of separation behavior must already be made *inside* the household.
- As doorstep campaigning is a rather costly intervention technique for enhancement of on-site separation of household waste, it could be questioned whether these investments are wise due to the small differences between households receiving and not receiving oral information in the case study area.
- Regarding monitoring: Monitoring should be made over a longer period after an intervention in order to reflect potential initial, but short-lived, changes. Only waste composition analyses give information on both the ratio of separately collected food waste in relation to the total food waste generation, as well as the quality in separately collected food waste. However, they do not give answers to the question *why* households behave in a certain way. Self-reported participation rates in recycling activities can diverge largely from levels estimated through waste composition analyses—thus, several different evaluation methods might be needed in order to assess the outcome of an introduced recycling scheme.
- Interventions should be scheduled to minimize risks of effects of seasonal changes, such as summer holidays. If households have adopted a routine to separate prior to such events, it is more likely that they will also continue after the holidays.

4.5.2.6 On-Site Separation of FOG

The Augustenborg area had previously experienced severe problems with clogging of the indoor plumbing system. Intense flushing has had to be performed on several occasions in the area, resulting in high costs for the property owner. The problems were, to a great extent, believed to be related to household use of FOG for frying. In order to decrease these problems, a system for separate collection of FOG was introduced in 2009. Households were asked to dispose of FOG waste in single-use plastic jars with lids (PE, 1.75 l), distributed among the households. Used jars were to be disposed of in the same recycling buildings where residents normally disposed all other waste. After disposal in recycling buildings, on-site separated FOG was collected in storage at the residence, and collected by the Municipal Waste Management Company on call from the facility manager for transportation to the treatment facility. In the treatment facility, FOG-jars were separated from their content in a pre-treatment plant (screw-press) currently used for separately collected food waste and packed food waste from industries, supermarkets, etc.

FOG was to be mixed with food waste for anaerobic digestion and collection jars for incineration with energy recovery. New jars could be collected from the recycling building in order to continue the on-site separation.

Written information regarding the recycling system was distributed to all households when the trial was started. Households also received oral information through a doorstep campaign, and a first set of the designated plastic jars aimed at on-site separation was distributed at the same time. Written and oral information was delivered in Swedish. Both written and oral information material was based on three key messages:

- Why should FOG be on-site separated.
- How should the households perform the on-site separation.
- What happens after the collection of on-site separated FOG.

The information also included symbols and pictures with the aim of giving instructions regarding FOG separation also in a language-neutral form. Thus, the information provided was both factual, with a focus on the practical arrangements regarding the recycling, and of a more persuasive nature focusing on the environmental benefits from recycling, earlier distinguished by de Young et al. (1993). Prepared information material was also created to be clear and easy to understand. The need for clear symbols and use of colors in the information directed toward households was emphasized, as Swedish was not the mother tongue to a large part of the households in the area. All information material was produced through close collaboration between the property owner, the Municipal Waste Management Company and the waste treatment company. At the same time as the doorstep information campaign was performed, a first set of plastic jars for on-site separation of FOG was delivered to each household.

The quantity of separated FOG was determined on a weight basis. Results show that the amount of separated FOG reached 0.35 kg/household per year in the study area during the study period. It was seen that many of the households chose to use other materials for their on-site separation of FOG. Other types of plastic jars and bottles, as well as glass jars with metal lids, were used. Only 7.5 % of all distributed jars were returned by the households during the trial period. Costs were, to a great extent, connected to the oral information campaign. The costs per collected kg of FOG in this study reached 1.2 euro, which could be seen as high. The costs were, to a very great extent, related to the information campaign performed. Thus, costs could be cut if choosing a different information strategy. Also, if the systems can result in a reduced need for flushing sewers, these costs might be acceptable (Davidsson et al. 2010). It was seen that the absolute majority of the designated plastic jars distributed for FOG-collection were never returned to the property owner. These jars were probably used by households for other purposes, such as storage of food in freezers or similar. The environmental benefits related to the trial were also vastly decreased by the large amount of virgin new plastic material used per kg of FOG collected. During the trial period, 157 g of plastic jars were used per kg of FOG collected. Based on previous investigations of greenhouse gas emissions from plastic production, this is equivalent to 0.3–0.7 kg CO_2-eq/kg of FOG collected. If plastic jars are combusted

Fig. 4.22 Funnel for FOG
collection distributed
amongst households

after separation of collected FOG, a further 2.5 kg of CO_2-eq emission are caused per kg of plastic jar (RVF 2003). Thus, each kg of separately collected FOG resulted in emissions of 0.7–1.1 kg of CO_2. This can be compared with a saving of 1.8 kg arising from substitution of 6.8kWh gasoline through anaerobic digestion of separately collected FOG (Davidsson et al. 2010). As collection, pre-treatment, anaerobic digestion and upgrading of FOG will also give rise to GHG-emissions, net savings from the system were very small. However, this comparison does not include emissions from flushing indoor plumbing systems with hot water, which had previously been done to avoid clogging.

Due to high costs and the reduced benefits from a climate perspective, the system was changed in 2012. Instead of the plastic jars, a funnel was distributed to each household. An inscription was placed on the funnel so that households could clearly read and understand what it should be used for. Although the new system decreases the energy and resource input for the separation of household FOG, it is still not optimal, as it means that plastic jars/bottles used by the household for the separate collection of FOG not will be recycled (Fig. 4.22).

4.5.2.7 Mobile Collection Points for Bulky Waste

Each month, households at the study site were given the opportunity to dispose of bulky waste in a mobile container, located at a central site. The waste was separated into bulky e-waste, bulky metal waste and other bulky waste, each type being collected separately. The container was operated by facility managers employed by the property owner and open for 3 hours (15:00–18:00) on the last Tuesday of each month. In connection to the collection, a swapping corner for bulky waste was established to make it possible for the residents to let old furniture and other items change owner instead of being thrown away.

The generation of bulky waste reached an average of 4.5 kg per household per month throughout an evaluation period of 2 years. Of these, only 23 % were disposed of in mobile recycling containers. Thus, more than 80 % of the bulky waste collected in the area was still found in recycling buildings, in basements and in the open air. One of the reasons for this is likely to be that many households do not want to wait for the next monthly collection occasion, that the opening hours of the container do not coincide with working hours or conflict with other activities, or that the distance to the collection point was seen as long. To reduce the risk of the latter, facility managers offered households use of trolley free of charge for transportation of bulky waste from residential buildings to the collection point.

The collected bulky waste mostly consisted of furniture (82.4 %), but also household items (8.7 %) such as carpets, textiles, kitchen equipment, etc., demolition waste (5.6 %) and waste related to leisure and sport (3.3 %) such as suitcases, bicycles and skis (Junghard 2011).

It was observed that the "swapping corner" was only used to a minor extent in the area. Although several items were set aside so that other households could pick them up, most of the items were later disposed of in containers, together with other bulky waste, by the end of the day, as the interest from other households was low. Thus, the idea of a swapping corner did not result in the potential waste minimization hoped for. This could be related to the fact that households only had a few hours to see and collect items disposed of by other households and a development of the swapping corner could results in improved results.

On several occasions, the facility managers running the mobile recycling point invited residents for a coffee and took the opportunity for an informal chat about the practical arrangements of waste handling and other practicalities in the area. Thus, the collection point was turned into a point for social interaction between households in the area, as well as between the property owner and residents.

4.5.2.8 Potential for Food Waste Minimization

Previous studies have estimated the amount of food waste generated in Swedish households at 72–100 kg/person per year (SMED 2011; Linné et al. 2008). Food production is commonly connected to major environmental impacts. According to the Swedish Environmental Protection Agency (SEPA), the production of the amount of food waste generated in Sweden every year corresponds to emissions of greenhouse gases equivalent to 2 million tons, or more than 220 kg CO_2-eq per capita per year (SEPA 2011). Thus, great environmental gains can be connected to food waste minimization, and the SEPA has selected food waste as a prioritized area for waste minimization. In order to estimate potential for minimization of household food waste, the composition of generated food waste must be investigated in further detail. Thus, an addition was made to the methodology for food

Table 4.7 Main and subfractions used in performed waste composition analyses

1. Avoidable	2. Unavoidable
1.1 Unopened food packaging	2.1 Tea leaves and coffee grounds, bones, peels, trimmings
1.2 Opened food packaging	2.2 Other unavoidable organic (e.g., flowers, tissues, etc.)
1.3 Other avoidable food waste (half-eaten food/prepared food)	2.3 Paper bags used for separate collection of food waste

Fig. 4.23 Examples of avoidable food waste observed in waste composition analyses. *Photo Anna Bernstad*

waste composition analyses used in the Augustenborg project, in the form of a differentiation between avoidable and unavoidable food waste. Both separately collected food waste and food waste disposed of in residual waste was divided into the two fractions, avoidable and unavoidable food waste. These two fractions were divided into a further six subcategories (Table 4.7).

Three waste composition analyses were performed where this division between avoidable and unavoidable food waste was used (Fig. 4.23). The results show that the amount of avoidable food waste reached 40 % on average. The average amount of avoidable food waste equals 64 kg/household per year. It is seen that the absolute majority of packaged (opened and unopened) food waste is not sorted out as food waste, but found in residual waste (Fig. 4.24).

Fig. 4.24 Unavoidable and avoidable food waste in residual waste (*left*), in separately collected food waste (*mid*), and the total (*right*), divided into subcategories

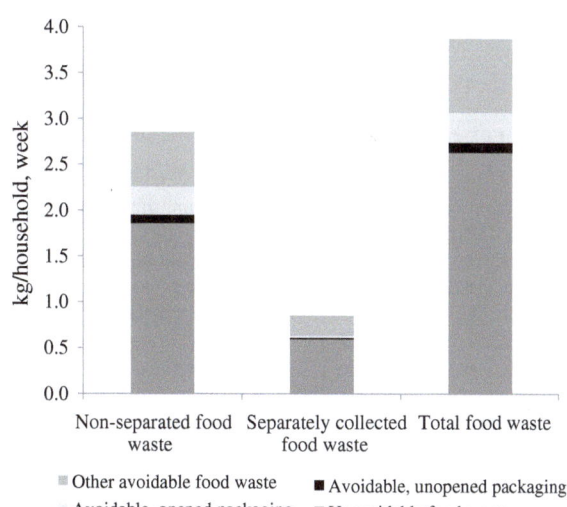

4.6 Aims, Methods and Outcomes

Although the aims in the two case studies described above were, to a great extent, similar, the methods chosen for achieving them differed greatly. Although the Bo01 project initially contained a substantial focus on community participation, decentralized solid waste treatment and recovery, as well as extensive use of information for behavioral change amongst households and visitors to the area, in the implementation stage, more focus was given to the technical systems. Several factors may have contributed to this; a will to "hide" solid waste management and thus use subterranean systems, a will to create a dense neighborhood and minimize use of the land in the area for purposes with low economic return (i.e., residential houses and offices) rather than waste collection and treatment facilities, as well as a lack of funding for some of the proposed activities. The outcome of this was a strong focus on hi-tech solutions, where technology was supposed to solve the waste problem. In the case of Augustenborg, the technologies used were simple, and, instead, the households were seen as key actors, and information and communication had a strong focus in the project. Opposite to the case of the Bo01 area, solid waste management had high visibility in the area, and recycling buildings and devices used for separate collection of waste in households are assumed to serve a pedagogic purpose as constant reminders of the importance of household contribution to sustainable solid waste management.

The case studies provided in this work show the introduction of novel technologies fokr solid waste management—commonly with a belief that these would be more attractive from an environmental, economic and social (behavioral) point of view, compared to previously tested systems. However, in-depth evaluations of the systems show that often some problems are solved while new ones are created. The case studies also show that communication activities toward households can be connected to several difficulties and not always result in the desired results. Thus, the search for optimal systems for municipal solid waste management is an iterative process we have not seen the end of yet.

References

Ando A, Gosselin A (2005) Recycling in multifamily dwellings: does convenience matter? Econ Inq 43(2):426–438

Bakas I, Clavreul J, Bernstad A, Niero M, Gentil E, Laurent A (2012) LCA applied to solid waste management systems: a comprehensive review. In: Oral session presented at: SETAC Europe 18th LCA case study symposium—sustainability assessment in the 21st century, 26–28 Nov 2012, Copenhagen, Denmark

Bernstad A, Davidsson Å, Tsai J, Persson E, Bissmont M, la Cour Jansen J (2012a) Tank-connected food waste disposer systems—current status and potential improvements. Waste Manag 33(1):93–103

Bernstad A, la Cour Jansen J, Aspegren H (2010) Property-close source separation of hazardous waste and waste electrical and electronic equipment—a Swedish case study. Waste Manag 31:536–543

Bernstad A, la Cour Jansen J, Aspegren H (2011) Life cycle assessment of a household solid waste source separation program: a Swedish case study. Waste Manag Res 29(10):1027–1042

Bernstad A, la Cour Jansen J, Aspegren H (2012b) Local strategies for efficient management of solid household waste—the full-scale Augustenborg experiment. Waste Manag Res 30(2):200–212

Bernstad A, Malmqvist L, Truedsson C, la Cour Jansen J (2013a) Need for improvements in physical pre-treatment of source-separated household food waste. Waste Manag 33:746–754

Bernstad A, la Cour Jansen J, Aspegren H (2013b) Door-stepping as a strategy for improved food waste recycling behaviour—evaluation of a full-scale experiment. Resour Conserv Recycl 73:94–103

Bolzonella D, Pavan P, Battistoni P, Cecchi F (2003) The under sink garbage grinder: a friendly technology for the environment. Environ Technol 24(3):349–359

Bjarnadottir HJ, Fridriksson GB, Johnsen T, Sletnes H (2002) Guidelines for the use of LCA in the waste management sector. Nordtest Report TR 517, Nordtest, Espoo, Finland

Dahlén L (2008) Household waste collection: factors and variations. Doctoral thesis Luleå University of Technology, p 33

Dalemo M, Sonesson U, Bjorklund A, Mingarini K, Frostell B, Jonsson H, Nybrant T, Sundqvist J-O, Thyselius L (1997) ORWARE—a simulation model for organic waste handling systems. Part 1: Model description. Resour Conserv Recycl 21:17–37

Davidsson Å, Bernstad A, Aspegren H, La Cour Jansen J (2010) Assessment of biogas production from source-separated fat, oils and grease (FOG) from households. In: Proceedings from third international symposium on energy from biomass and waste. Venice, Italy, 8–11 Nov 2010

de Young R (1993) Changing behavior and making it stick. The conceptualization and management of conservation behavior. Environ Behav 25:485–505

Derksen L, Gartrell J (1993) The social context of recycling. Am Sociol Rev 58:434–442

Elforsk (2008) Miljövärdering av el—med fokus på utsläpp av koldioxid. Elforsk AB, Stockholm, Sweden

Esaiasson P, Gilljam M, Oscarsson H, Wängnerud L (2007) Metodpraktikan. Nordstedts Juridik AB, Stockholm

Ewert S, Henriksson G, Åkesson L (2009) Osäker eller nöjd—kulturella aspekter på vardagens avfallspraktik. KTH Arkitektur och Samhällsbyggnad. ISSN 1652-5442 TRITA-INFRA-FMS 2009: 3. Stockholm, Sweden

Gentil EC, Aoustin E, Christensen TH (2009) Greenhouse gas accounting and waste management. Waste Manag Res 27:696–706

Gruvberger C, Aspegren H, Andersson B, la Cour Jansen J (2003) Sustainability concept for a newly built urban area in Malmö, Sweden. Water Sci Technol 47(7–8):33–39

Hoppe M (2013) Waste management technology in everyday life. Final report, Jan 2012. Master student Thesis in Applied Cultural Analysis at Lund University

JRC (2011a) Support JRC scientific and technical reports. Supporting environmentally sound decisions for waste management: A technical guide to life cycle thinking (LCT) and life cycle assessment (LCA) for waste experts and LCA practitioners

JRC (2011b) Support JRC scientific and technical reports. Supporting environmentally sound decisions for food waste management—a technical to life cycle thinking (LCT) and life cycle assessment (LCA) for waste experts and LCA practitioners

JRC (2011c) Supporting environmentally sound decisions for construction and demolition (C&D) waste management http://lct.jrc.ec.europa.eu/pdf-directory/D4B-Guide-to-LCTLCA-for-C-D-waste-management-Final-ONLINE.pdf

Junghard E (2011) Collection and reuse potential of household bulky waste—a study on bulky waste management in Augustenborg, Master Thesis, Waste and Environmental Engineering, Lund University

Kirkeby JT, Birgisdottir H, Lund Hansen T, Christensen TH, Bhander GS, Hauschild M (2006) Evaluation of environmental impacts from municipal solid waste management in the municipality of Aarhus, Denmark (EASEWASTE). Waste Manag Res 24:16–26

la Cour Jansen J, Christensen T, Davidsson Å, Lund Hansen T, Jönsson H, Kirkeby J, (2007) Biowaste—decision support tool for collection and treatment of source-sorted organic municipal solid waste. TemaNord 2007:602. Nordic Council of Ministers, Copenhagen

Linné M, Ekstrandh A, Englesson R, Persson E, Björnsson L, Lantz M (2008) Den svenska biogaspotentialen från inhemska restprodukter/The Swedish biogas production from internal waste (in Swedish). BioMil AB/Envirum AB, Lund

Malmberg J, Nilsson C (2011) Hyresgästperspektiv på matavfallsinsamling i Augustenborg. www.sysav.se/upload/.../Hyresgästperspektiv%20på%20matavfall.pdf

Marashlian N, El-Fadel M (2005) The effect of food waste disposers on municipal waste and wastewater management. Waste Manag Res 23(20):20–31

Martin M, Williams ID, Clark M (2006) Social, cultural and structural influences on household waste recycling: a case study. Resour Conserv Recycl 48(4):357–395

Mathiesen BV, Münster M, Fruergaard T (2009) Uncertainties related to the identification of the marginal energy technology in consequential life cycle assessments. J Cleaner Prod 17:1331–1338

McDonald S, Ball R (1998) Public participation in plastics recycling schemes. Resour Conserv Recycl 22(4):123–141

McDougall FR, White P, Francke M, Hindle P (2001) Integrated solid waste management—a lifecycle inventory, 2nd edn. Procter and Gamble. Blackwell Science Ltd., London

Mee N, Clewes D, Phillips P, Read A (2004) Effective implementation of a marketing communications strategy for curbside recycling: a case study from Rushcliffe, UK. Resour Conserv Recycl 42:1–26

Michelson W (1990) Measuring macroenvironment and behaviour: the time budget and the time geography. In: Bechtel RB, Marans RW, Michelson W (eds) Methods in environmental and behavioural research. Van Nostrand Reinhold, New York, pp 216–243

Miranda ML, Aldy J (1994) Market-based incentives and residential solid waste. J Policy Anal Manag 13:681–698

Nilsson P, Hallin P-O, Johansson J, Karlén L, Lilja G, Pettersson BÅ, Pettersson J (1990) Källsortering med avfallskvarnar: En fallstudie i Staffanstorp (Source-separation with food waste disposers: a case study from Staffanstorp) (in Swedish). Report REFORSK FoU number 54, Lund, Sweden

Nordtest (1995) Solid waste, municipal: sampling and characterization. (Nordtest method NT ENVIR 001, Finland. www.nordtest.org. Accessed 11 Feb 2013

Parfitt JP, Flowerdew R (1997) Methodological problems in the generation of household waste statistics—an analysis of the United Kingdom's national household waste analysis program. Appl Geogr 17:231–244

PCR (2008) Product category rules (PCR) for preparing an environmental product declaration (EPD) for solid waste disposal services. PCR 2008:02. http://www.environdec.com

Petersen C (2004) Conditions and constraints for waste management. Collection, characterisation and producer responsibility in Sweden, Doctoral thesis No. 2004:10 Dalarna University College and Chalmers University of Technology, Borlänge and Gothenburg, Sweden

Pitard FF (1993) Pierre Gy's sampling theory and sampling practice: recycling intensity. Environ Behav 31(2):267–290

Raunkjaer K, Hvitved-Jacobsen T, Nielsen PH (1995) Transformation of organic matter in a gravity sewer. Water Environ Res 67:181–188

RVF (2003) Förbränning av avfall Utsläpp av växthusgaser jämfört med annan avfallsbehandling och annan energiproduktion. http://www.avfallsverige.se/fileadmin/uploads/Rapporter/ Utveckling/Rapporter%202003/U2003-12.pdf

SEPA (2007) Materialströmmar—ett bättre sätt att samla in hushållsavfall? Utredning av förutsättningar för insamling och återvinning av hushållens avfall i materialströmmar/Material streams—a better way to collect household waste? Investigation of the prerequisites for collection and recycling of household waste in material streams (in Swedish). Stockholm, Sweden

SEPA (2011) Konsekvensutredning av en minskning av matavfallet med 20 % mellan 2010–2015. Swedish Environmental Protection Agency, Stockholm, Sweden

SMED (2011) Matavfall 2010 från jord till bord. SMED Rapport Nr 99 2011. Jensen C, Stenmarck Å, Sörme L, Dunsö O. Naturvårdsverket, Stockholm, Sweden

Stirling A (1997) Limits to the value of external costs. Energy Policy 25:517–540

Sundqvist J-O (1999) Life cycles assessments and solid waste—guidelines for solid waste treatment and disposal in LCA. IVL, Swedish Environmental Research Institute. Report 279. AFN, Swedish Environmental Protection Agency, Stockholm, Sweden

Timlett R, Williams ID (2009) The impact of transient populations on recycling behaviour in a densely populated urban environment. Resour Conserv Recycl 53:498–506

White P, Francke M, Hindle P (1995) Integrated solid waste management—a life-cycle inventory. Chapman and Hall, London

Åberg H (2004) Boendeperspektiv på hushållsavfall och på system för insamling och behandling i Västra Hamnen, Malmö, Report 37. Department of Home Economics, University of Gothenburg, ISSN 1403-7033

Chapter 5
Collaboration Outcomes

Keywords Technology development · Methodological development · Organizational development · Decision support tool

5.1 Technological Development Through External Evaluation

One of the aims of the projects presented above has been to introduce and evaluate new technologies for household waste disposal, collection, and further treatment. Several new technologies were introduced at full scale in Malmö. Regardless of the often great knowledge and long experience among technicians in the local waste management authority, there is commonly a lack of time for systematic evaluation. Thus, the external evaluation on the part of the university resulted in the thorough identification of weak spots—both from a strictly technological point of view, and, principally, when technologies were studied from a systems perspective. Thus, the collaboration with the university resulted in the development of these technologies and systems.

- *Tank-connected food waste disposal systems*
  As evaluations indicated high losses of organic matter and nutrients from the tanks, changes were made in the design of the tank when introduced in new developments.
- *Improved infrastructure for on-site separation inside the household*
  Development of an on-site separation concept with the aim of facilitating primarily food waste separation, but also separation of recyclable packaging, acknowledging that behavioral changes in relation to recycling already had started in the household (Fig. 5.1).

A. Bernstad Saraiva Schott et al., *Modern Solid Waste Management in Practice*,   69
SpringerBriefs in Applied Sciences and Technology, DOI: 10.1007/978-1-4471-6263-6_5,
© The Author(s) 2013

Fig. 5.1 On-site separation
concept developed within the
project. *Photo* Gugge
Zelander

- *Cabinets and cages for separate collection of hazardous waste and e-waste*
  Through the evaluations provided by the university, cabinets for hazardous
  waste were evaluated in a full-scale environment. Problems with safety and
  vandalism, such as children removing and crushing light bulbs collected in open
  boxes on the sides of the cabinets, were recognized during the trial period.
  Changes in the design of the cabinets could be made before the same system was
  introduced in new areas in order to avoid this in the future.
- *System for collection of FOG from households*
  The system for separate collection of FOG introduced in the Augustenborg
  study area, was considered successful in relation to the system users (i.e.,
  households), but expensive and results in large environmental impacts. Thus, the
  use of disposable plastic jars for separate collection was changed to a system
  with collection of FOG in used plastic packaging, decreasing both environmental impacts and costs.
- *Improving pre-treatment of food waste with screw press in order to reduce
  losses of biodegradable material and nutrients*
  System studies showed that the potentially large losses of biodegradable
  material in the pretreatment of household food waste can reduce overall
  potential benefits from the system. Different strategies for reduction of refuse
  could be tested at the pretreatment facility. However, as systems studies also
  showed the great importance of maintaining high quality in produced biomass,
  all evaluations of introduced changes were made starting from the point that the
  quality should not be affected.

5.2 Methodological Development

In several cases, new methodologies were needed in order to evaluate the effect of
introduced technological and collection systems. In many cases, the practical
arrangements in the full-scale systems did not facilitate systematic evaluation of

the systems in question. Thus, another point learned from the project is the need for collaboration between the designer of the system and the evaluating agent prior to installation in order to facilitate future evaluations. This was clearly put into practice, among others, in the case of further installations of tank-connected food waste disposers in new areas of the city, where the university was actively participating in the design of the new systems. Another example was the need for manuals for performance of waste composition analyses, where a difference is made between avoidable and nonavoidable food waste in order to estimate the potential for environmentally relevant significant food waste minimization.

5.3 Contact Between Engineering and Users: Every Link of the System is Important

In the case of the Bo01 project, the contact between academia and different divisions of the municipality (one of them being the waste management section, at that time, part of the Department of Parks and Public Spaces), private waste collection entrepreneurs as well as property owners, was not formalized. This might have been one of the reasons why much of the knowledge learnt from the academic evaluations of the waste management systems in the area—both from a technical perspective, as well as from the perspective of user-friendliness—were not applied in later projects in the same area, such as the installation of food waste disposers in a high-rise building in the area in 2007. This also shows the importance of documentation, dissemination, and internalization of research results within the structures responsible for decision-making for the organization of solid waste management.

The organization of the Augustenborg project, i.e., the close consistent contact with system users through different methods provided a quite uncommon contact between the users and the designers of waste disposal systems, in which the design could be adjusted in order to meet the needs of the households. The holistic approach taken in the project resulted in close contact between the designers of the technical infrastructure used in the waste management systems and the actual users of the systems. In many cases, it became clear that issues not commonly prioritized from a technical point of view and in the design of the system, were of great importance for the users of the same system. Several examples of this were seen throughout the projects:

- Much focus was initially directed to innovative waste collection technology, while little attention was paid to the infrastructure for on-site separation inside households. For the households, the lack of space inside the home created a barrier and increased reluctance to participation.
- While inlet doors in vacuum system chutes were dimensioned to minimize the risk of waste clogging the underground tubes, this increased the need for

households to visit recycling centers in order to dispose of waste that did not fit into the chutes.

- While great effort is put into making food waste disposers that grind the matter into particles that are as fine as possible in order not to cause problems in the sewage system, one of the largest obstacles for households was, according to the interviews performed, that the actual grinding was too time-consuming—especially when larger amounts of food waste were to be disposed of. Thus, there is a risk that large parts of the potential substrates for biogas are disposed of together with residual waste if the grinders are not developed to reduce the timely grinding process.

The project clearly showed that waste management systems relying on on-site separation must, to a greater extent, consider the convenience for households to participate in recycling schemes. High efficiency in further treatment can commonly not compensate for a low participation on the part of the households.

5.4 Organizational Development

The integration of solid waste management infrastructure in the urban environment calls for increased collaboration among different municipal departments and private interests. An example is the installation of tank-connected food waste disposers. These are a part of the waste management system and thus under the responsibility of the waste management department. However, the Department for Land Tenure and Use was the owner of the land in which the tank was to be placed. The Department for Urban Planning had to make room for the tank in the area plan. The Department for Parks and Public Spaces was also important in this process, as they were responsible for management of public spaces and thus the area where tanks were to be placed. Also, the connection of solid waste to the sewage system called for collaboration with the VA SYD wastewater division. Thus, several different public departments had to collaborate in the development of these installations. In order to coordinate needed actions, and based on the experience from the Bo01 project, a project group was initiated by the VA SYD waste management department, consisting of relevant actors. However, at the time the project group was initiated, the planning process of the area had already been finalized and large parts of the sewage systems were already in place. Thus, due to the fact that the process was initiated much too late, the project became significantly more expensive than needed. Also, the placement of the tanks was suboptimized by the fact that these installations had not been considered in the planning process. Thus, although the forming of a cross-sectorial project group was a step forward, this is a clear example of how solid waste management must be considered early in the planning process for increased sustainability.

5.5 Scientific Basis for Policy Making

As previously stated, local policy making within the solid waste management area is a result of several different parameters, including national and supranational legislation and objectives, the economy, and several local factors. However, decisions taken on a daily basis within the local waste management authority is commonly not based on scientific grounds. Due to the close collaboration between the university and Malmö, the local waste management authority, as well as local politicians, could use results based on scientific works as the basis for their decision-making. Thus, the local waste management authority could, in this case, justify their decisions in relation to strategies, and make investments with scientific results showing the potential environmental benefits related to strategies undertaken by the municipality. This has strengthened civil servants in the municipality in their communication with households, as well as with local politicians. As a clear result, outcomes from the projects showing that separate collection of household food waste for later biogas production from a climate perspective is beneficial in relation to both incineration and decentralized composting, serving as an important input in the process of introducing a mandatory separate collection of household food waste in the municipality in 2012.

The use of life-cycle assessment as a decision support tool in Malmö unmistakably showed the potential for this method in achieving sustainable waste management. The tool clearly showed increased environmental benefits related to some treatment alternatives compared to others, but also identified processes that are of greater and lesser importance in relation to overall results. Thus, the tool could be used to prioritize efforts on the part of the municipality. The results also clearly state the importance of including several different sectors of society in the discussion on sustainable waste management.

Methods used in the evaluation of waste management projects performed in the municipality of Malmö were also used to investigate the potential effects of changes in national waste management policy-making, such as a change from material recycling based on packaging to a material type based system. The combination between different tools for evaluation could show that, even though such a change would not have a very great effect on the total amount of potentially recyclable materials from households, it could result in important environmental benefits, as the material fractions where a change would have the largest impact also are the fractions connected to the largest positive environmental impacts, if recycled. This shows the importance of holistic approaches in evaluations of waste management policy-making.

Although life-cycle assessment is a recognized methodology for systems analyses and assessment of environmental impacts, the results presented in the projects performed in collaboration with the municipality of Malmö showed that results can differ depending on assumptions made by the LCA practitioner, especially in relation to the environmental profile of used and substituted energy bearers. The differentiated interest of actors involved in the waste can result in a willingness to depart from different assumptions in the life-cycle assessment. Thus, in spite of close collaboration between the different actors involved in the waste management treatment chain and academia, it is of extreme importance that the academic sphere maintains independence in presentation and interpretation of the results obtained.

5.6 Use of Test Beds as a Method in Development Work

Both the Bo01 and the Augustenborg projects are examples of how the city has chosen to test new systems for waste management on a smaller scale. Using these areas as test beds for innovative techniques and strategies was seen as a fruitful strategy in relation to the aim to strive for constant development and improvement of the systems and strategies used in the city. Due to the often large investment costs within the area of solid waste management systems, starting small and testing different systems in parallel is essential, as it allows for mistakes to be made on a smaller scale prior to going citywide. The test bed concept also facilitates evaluation of introduced technologies and strategies, resulting in improved decision support, and can be seen as a key factor in order to avoid building the cities of the future on the knowledge of the past.

On-site separation is viewed by Malmö as essential in sustainable waste management. Thus, the human behavior is a key factor when seeking to increase sustainability. Many challenges still remain in order to increase efficiency in waste treatment technologies, but potential gains in such improvements commonly may not compensate for losses earlier in the system, related to inefficient on-site segregation separation. This was clearly shown by life-cycle assessments performed in collaboration with the university. Thus, the collaboration among academia has resulted in a decreased focus on technological solutions and an increased focus on behavior. These issues had not been neglected previously, but the collaboration has given the Department for Solid Waste Management tools for analyzing these within a scientific framework.

Chapter 6
New Projects, Building on Previous Experience

Keywords Esthetic waste management · Energy balance · Waste-to-energy

In the years after the Bo01 project, new developments were made in the Western Harbor area in Malmö (Fig. 6.1). A large development site was also built in the Hyllie area, south of the city center (Fig. 6.2). Much of the experience gathered in relation to solid waste management from previous development areas came to be used in the development of these two sites.

6.1 Fullriggaren: Further Development of the Grinder Concept

Much knowledge learnt from the Bo01 project came into use in the development of *Fullriggaren*—a new development in the Western Harbor with 624 apartments, a kindergarten, indoor parking, and offices, in which 13 different developers were involved (Fig. 6.3). A dialogue development process was introduced—similar to the one used in the development of the Bo01 project 10 years earlier. As this development also took place on municipally owned land, in this case too, environmental requirements were imposed by the municipality in order to ensure certain desired standards were met in the development of the area. A "Sustainability Agreement" was reached as a part of the dialogue process. On-site separation of packaging waste was mandatory and had to be provided for all households in the area, but no solutions for other fractions, such as e-waste, bulky waste, or hazardous waste, were discussed in the agreement. Due to the compact architectural design of the area, it was decided that tank-connected food waste disposers should be used.

A number of points were learnt through the external evaluation of the tank-connected food waste disposal system previously installed in the Western Harbor.

A. Bernstad Saraiva Schott et al., *Modern Solid Waste Management in Practice*,
SpringerBriefs in Applied Sciences and Technology, DOI: 10.1007/978-1-4471-6263-6_6,
© The Author(s) 2013

Fig. 6.1 Overview of Fullriggaren in the western harbor area. *Photo* Malmö city urban planning office

Fig. 6.2 Overview of Hyllie Allé. *Photo* Malmö city urban planning office

Thus, two different types of tanks were installed in the area—one very similar to the ones previously installed in the Western Harbor and one slightly modified.

As pipes from several different facility owners were to be connected to the same tank, a clear division of ownership and responsibilities was needed. It was decided

Fig. 6.3 Fullriggaren from above with the buildings constructed by 13 different developers marked (*left*) and with the two separate systems for food waste disposers (*right*). Illustration: VA SYD

that facility owners were owners of the double piping system and pumps on private land, while the municipality was responsible for maintenance of the pipes laid on public land and both tanks. Facility owners were to pay a monthly fee for collection and treatment of food waste sludge from the tanks. The contract period between facility owners connected to the tank system and the municipality was set at 4 years. The evaluation period was set at 2 years after installation. During this period, no collection fees were to be charged by the municipality. Also, a framework for evaluation of the system in relation to several different aspects was developed:

- Technical performance of the systems—Comparison of the different tank designs, made in collaboration with the Water and Environmental Engineering Faculty at Lund University.
- Household behavior—Assessment of on-site separation behavior and user-friendliness, made in collaboration with the Department of Water and Environmental Engineering and the Ethnology Department at Lund University.
- Organization, process, and economy—Assessment of the organizational form of the process of introducing food waste disposers in the area as well as costs of installation, use, and maintenance, made in collaboration with Malmö University.

6.2 Hyllie: Increasing Integration of Solid Waste Management in the Urban Planning Process

Hyllie is a new development area in southern Malmö. The area has been planned for 9,000 apartments, 120,000 m^2 of office space, more than 100,000 m^2 for commerce, a new sports arena, fair, and shopping center (Fig. 6.4). Sustainability

Fig. 6.4 An overview of the developments in Hyllie in 2012. *Photo* Malmö city urban planning office

has been one of the key words in the development process. Two separate structures have been introduced in the process with the aim of increasing the environmental sustainability in the area.

As in the planning process of Fullriggaren, a "Sustainability Agreement" was made between developers active in the area. However, due to the larger size of the Hyllie area compared to Fullriggaren, a smaller area, with a total of 1,500–2,000 apartments, was selected as a pilot area. Also, different from Fullriggaren, the Hyllie Agreement contained much more detailed criteria and objectives for solid waste management:

- Plan and shape public areas so that waste sorting is facilitated for visitors to the area
- Plan waste management so that noise and transportation are minimized
- Place a local recycling center in the area
- Design buildings and waste management systems to facilitate waste sorting into at least nine different fractions. Systems for waste sorting and recycling should be accessible and simple to use by the households. Spaces should be provided for waste sorting and recycling both inside households as well as other parts of the building and area
- Separate collection of household waste should be provided within the facility or in close proximity to it
- Waste management areas should be designed to provide a certain flexibility through the provision of at least 10 % extra space for waste management in order to introduce possibilities of waste separation into more fractions in the future
- Waste minimization and waste sorting for potential recycling is to be a focus area during the construction of buildings in the area. The amount of landfilled waste from each building site should not exceed 12 % of the mass. Total waste production should not exceed 15 kg/m^2 constructed area (heated as well as nonheated).

Recycling targets were established for all waste fractions included in the producer responsibility legislation as well as for household food waste. These targets are to be evaluated through waste composition analyses performed by the Department for Waste Management, and feedback to the residents in the area based on the results from the analyses is to be distributed through coordinated actions by the Department for Waste Management and the Municipal Department for Environmental Protection. The total waste production, as well as specification of production within different waste categories, is to be reported to the municipality by the final inspection of each building in the area.

As the criteria relate both to waste management in public spaces, waste from privately owned facilities, use of public and private areas for construction of recycling facilities, several different actors, both public and private, would have to be active in order to fulfill the objectives in the Sustainability Agreement.

Initially, different solutions for food waste collection were discussed in the process. Two systems supported by the Department for Public Spaces were vacuum systems and food waste grinders, directly connected to the sewage system. The main benefit of these solutions was a reduction of the amount of waste visually present in the area. Both vacuum systems and sewage-connected food waste grinders were, in this context, seen as a quick and an easy solution for one of the largest waste fractions, and they would reduce the visual impact of waste in the area. However, experiences from the Bo01 area did not support the vacuum system, and large parts of the Department for Waste Water Treatment did not support the sewage-connected food waste grinder system.

Due to many still unanswered questions in relation to impacts on the sewage system, the wastewater treatment plant, etc., the Department of Solid Waste Management saw a need for testing and evaluation. A proposal for a 2-year trial period was made to the developers. Thus, developers saw a risk in investing in grinders that would have to be removed after 2 years in case of negative results from the evaluation. Also, evaluations from the two earlier testing sites with food waste grinders in the city had shown that they did not result in the improved standards for households previously presumed (see Hoppe 2012). Thus, a previously assumed increased property value and selling argument related to food waste grinders was now much inferior than previously assumed.

Thus, another strategy was chosen. If the approach in the Bo01 project had been to hide waste in subterranean systems, the tactic in Hyllie was instead to make the possibilities for sustainable waste behavior and management visual in the area. The Department for Waste Management was now designated the task of developing a more esthetic and attractive recycling building concept for the households in the area, with the focus on instantaneous feedback to the user, personalized statistics, interactive information, and visualization of environmental benefits. The design of the recycling building will build on experiences gained in previous waste management projects performed in the city as well as on knowledge gained through collaboration with researchers in the areas of sociology and behavioral science.

The producer responsibility legislation on packaging and newspapers has put a focus on on-site separation of these fractions in Sweden. Thus, waste separation has had a strong recycling focus. However, from the perspective of reducing the toxicity of generated waste, another of the Swedish Environmental Objectives, separate collection of fractions, such as hazardous waste, batteries, and e-waste is of even greater importance. Although Sweden is a leading country in relation to separate collection and treatment of e-waste, and several systems for separate collection of hazardous waste are currently in place in the country, previous studies have shown that these types of waste are still commonly found in Swedish household residual waste (Swedish Waste Management Organization 2008). Another waste fraction that is quite commonly forgotten in Swedish municipal waste management planning is bulky waste. The most common system for collection of such waste today is transportation to recycling centers, which are often located on the outskirts of larger cities. However, this collection system generally results in the use of private cars for transportation of bulky waste to the disposal center, which is not in line with the sustainability goals in the city in the area of transportation, where several objectives exist with a focus on increasing travel by bicycle and public means of transport, while decreasing the use of private cars (City of Malmö 2011).

Thus, new forms for collection of hazardous waste, e-waste, and bulky waste was sought as a part of the Sustainability Agreement in the Hyllie Allé area. The initial plan was a mini recycling center for these waste fractions. The center would be financed by the different facility owners in the area, and was seen as a service for the residents in the area. The plan was to fit such a center into a multistorey car park in the area. However, it was soon realized that this would be more difficult than initially thought. As facility owners were to pay for the system, they wanted to make sure that only residents in the area made use of the center. How could this be guaranteed? Instead, the concept of a monthly visit from the mobile recycling container also used in the Augustenborg area was chosen in the end. Also, due to the closeness to one of the existing public recycling centers in the city, it was suggested that the center could be extended with another entrance for bicycles, as a way to promote sustainable transportation of waste to the recycling center.

As a part of the sustainability context, a Climate Contract was established between the City of Malmö and the energy company, Eon, with the aim of providing the area with 100 % renewable or recycled energy by 2030.

A triple helix structure was developed in order to develop the conceptual framework and a tool for estimation of current and future energy balance in the area. The structure consisted of the energy company, the municipality (represented by the Municipal Department for Environmental Protection and VA SYD), and academia. An academic reference group was constructed with the aim of providing the project with realistic data on current and future processes of relevance to the energy balance throughout the area. As the contract covered several other areas apart from the waste sector, researchers from several different areas of relevance were involved, including energy efficiency and sustainable urban development, among others. In order to also grasp aspects related to behavior, researchers in the field of social anthropology were included in the reference group.

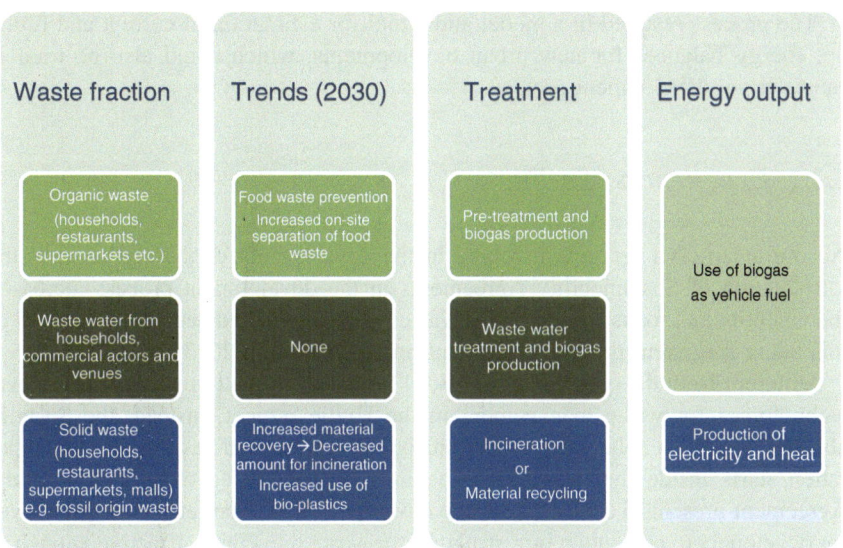

Fig. 6.5 Waste treatment processes considered in the climate contract, presenting acknowledged trends, and used waste-to-energy processes. It is to be noted that waste of fossil origin is not included in the balance, and material recycling does not result in any energy recovery

The management of solid waste, through anaerobic digestion of food waste and incineration of residual waste with heat and power recovery, had a central part in the contract. In order to predict the need for additional production of renewable energy to the Hyllie area (apart from the energy gained thorough waste-to-energy solutions), the energy company and the municipality had to estimate future potential for waste-to-energy from the area (Fig. 6.5). This included a prognosis of future waste amounts, on-site separation of food waste and recyclables, and increased use of bioplastics (which might not be recyclable, but also have a lower heating value).

In the conceptual framework for the energy balance, three different scenarios were constructed; a worst case, a realistic case, and a best case (Table 6.1). It should be highlighted that these scenarios were constructed based on what was assumed to be the best and worst from an environmental point of view.

Table 6.1 Assumptions used in the construction of a worst case, a realistic case, and a best case scenario for the on-site separation of food waste and dry recyclables in the energy balance for Hyllie

Waste fraction	Scenario	2015 (%)	2020 (%)	2030 (%)
Food waste	Worst case	25	30	40
	Realistic case	35	40	45
	Best case	40	60	100
Dry recyclables	Worst case	31	31	31
	Realistic case	40	45	55
	Best case	50	75	100

The process resulted in a model and a tool for estimation of current and future net energy balances for new urban developments, which could also be used in future urban development projects.

6.2.1 Experiences: So Far

According to civil servants involved in the development of the Hyllie area, the introduction of Sustainability Agreements in the development process increased the visibility and focus related to this waste management and decreased the risks of this being forgotten in the development process (Fossum 2013).

Different from the outcome of the Bo01 area, little attention was given to novel technologies for waste management in the Hyllie project. Instead, the Sustainability Agreement relates to behavioral issues and recognizes that on-site separation starts inside the household. The introduction of recycling targets in the Agreement presents a challenge for the developers. These targets create incentives for developers to present an infrastructure for waste recycling, attractive enough to result in behavior changes, and thus fulfillment of the targets.

In order to structure the follow-up of the Agreement, checklists were developed in relation to the different procurements and objectives agreed upon. Checklists contained questions related to each different area of the Sustainability Agreement. In relation to solid waste management, five questions were asked (Fig. 6.6).

The checklists were also used in meetings with the developers, where each developer presented the procurements and strategies used in their particular project. Through the follow-up, it becomes obvious if anyone was lagging behind in relation to any of the different areas in the agreement, solid waste management

Finalized (Yes/No)	Waste management	Finalized (date)	Comment
	Waste management has been planned to reduce noise and transportation		Describe how
	User-friendliness and accessibility have been considered in design of waste management solutions in the buildings as well as in apartments/commercial facilities.		Describe how
	Apartments and commercial facilites have been planned to facilitate separate collection of waste.		Describe how
	Future demands related to waste management have been considered in planning and design of waste solutions.		Describe how
	Reused materials have been used in the buildings.		Describe what

Fig. 6.6 Checklist for evaluation of criteria connected to waste management in the sustainability agreement in Hyllie

being one of them. This process is important, as no direct sanctions were connected to the agreements. A difference between the two agreements is that the one in Hyllie also contained behavioral parameters. Thus, in order to succeed, developers had to construct systems for separate household collection attractive enough to reach targets set for sorting rates.

The early integration of waste management in the Hyllie project was reached through inclusion of the Department for Waste Management in the development dialogue between developers and Malmö. Such dialogues had been held both in the development of Bo01 and *Fullriggaren* in order to establish criteria for development on municipally owned land. However, in those cases, the dialogue had been restricted to inclusion of representatives from the City Planning Office. In Hyllie, the waste management department was invited to several dialogue meetings to discuss waste management strategies for the area. In this way, solid waste management gained increased importance in the development and urban planning process. Waste management was also included in binding agreements between the municipality and the developers, which the municipality saw as essential. Thus, waste management in the Hyllie development project was no longer only an "add-on," but increasingly an integrated part of the development process.

The focus on a smaller part of the large development area in Hyllie in the form of a pilot project was seen as successful, as the dialogue process was reduced to a manageable number of actors. Within the coming years, a development of the remaining part of Hyllie will continue. In this process, an environmental strategy will be developed for the whole area. This strategy will have to be considered in the physical planning process. According to the municipality, this will increase the integration of solid waste management at a very early stage in the process (Fossum 2013). The strategy will also build on the experiences made in the smaller area, currently under development. In a wider perspective, the Climate Contract is also to be seen as a test of a planning tool which can be used in other development areas in Malmö in the aim of reaching the overall municipal goals of decreasing the total amount of energy used in the municipality with 40 % until 2030, and that 100 % of the energy used should be based on renewable sources by the same year (City of Malmö 2009).

References

City of Malmö (2009) Environmental program for the city of Malmö. http://malmo.se/download/18.76105f1c125780a6228800031254/Milj%C3%B6program+f%C3%B6r+Malm%C3%B6+stad+2009-2020.pdf

City of Malmö (2011) Malmö Stad Trafikmiljöprogram 2012–2017. http://www.malmo.se/Kommun–politik/Om-oss/Forvaltningar–bolag/Gatukontoret/Plan–atgards–och–policyprogram/Trafikmiljo.html

Fossum T (2013) Civil servant at Malmö City Planning office, Interview February 2013

Hoppe M (2012) Waste Management Technology in Everyday Life. VA SYD final project report, January 2012. Applied Cultural Analysis at Lund University

Swedish Waste Management Organization (2008) Vart tar smått el-avfall från hushåll vägen? Report 2008:03. Swedish Waste Management Association, Malmö

Chapter 7
Future Challenges

Keywords Waste management economy · Resource recovery · Household behavior · Collaboration forms

Several steps have been taken toward a more sustainable solid waste management in Malmö over the last few decades. However, several challenges still remain. Some of them are discussed here.

7.1 From Decentralized to Centralized: Maintaining the Pedagogic Link

In the transformation process of the Augustenborg residential area to an EcoCity in the late 1990s, ecological sustainability, at this point, was, to a great extent, still connected to eco-villages, and, thus, decentralization was seen as a key factor for increased sustainability. This was the reason behind the system for decentralized composting installed in the Augustenborg area. This concept was also markedly present in the initial planning of the Bo01 area. However, this view was abandoned, partly due to the fact that this would reduce the attractiveness of the area, but also through use of system analyses of different waste management solutions, showing that centralized biogas production was preferable to a decentralized alternative.

The same conclusion was drawn in life-cycle assessments performed in the Augustenborg project, as these showed clear advantages related to the use of household food waste for biogas production rather than decentralized composting. However, decentralized composts, like the ones previously used in the Augustenborg area, contribute a simple pedagogy for the households, where they can clearly see with their own eyes how the food waste becomes soil which can be used for cultivation. In the case of anaerobic digestion, households must use other material—collection bags, food waste needs to be transported and households can no longer, with their own eyes, see the transformation from waste to useful products.

A. Bernstad Saraiva Schott et al., *Modern Solid Waste Management in Practice*, SpringerBriefs in Applied Sciences and Technology, DOI: 10.1007/978-1-4471-6263-6_7, © The Author(s) 2013

Thus, although the anaerobic digestion is preferable from an environmental perspective, this is not always clear to households, and the municipality faces a great challenge in conveying this message to the households.

7.2 Behavioral Changes

On-site segregation separation is viewed by Malmö as essential in sustainable waste management. Thus, human behavior is a key factor when seeking to increase sustainability. Many challenges still remain in improving efficiency in waste treatment technologies, but potential gains from such improvements often may not compensate for losses earlier in the system, related to inefficient on-site segregation separation. This was clearly shown by life-cycle assessments performed in collaboration with the university. Thus, the collaboration with academia has resulted in a decreased focus on technological solutions and increased focus on behavior. These issues had not been neglected previously, but this collaboration has given the Department for Solid Waste Management tools for analyzing these within a scientific framework.

A commonly asked question in relation to behavioral changes and waste sorting is—how can we make people care? Information can be seen as a key in accomplishing behavioral changes, the reason being that once people know what they should do and why, they will do so. However, studies have shown that providing information not always result in increased knowledge, and that increased knowledge in many cases not result in changed behavior (Borgstede and Biel 2002). Information about waste management is today likely to drown in a huge amount of other messages, daily competing to attain our attention. Thus, finding information channels and messages which catches the inhabitants' attention is a challenge in itself. A particular challenge is also seen in relation to separate collection of household food waste, where municipalities at the same time are obliged to strive toward a minimization of food waste generation. Thus, setting goals and communicating a set amount of food waste that should be collected separately by the households is difficult—there is a risk in misinterpretation of such messages, as if throwing away food would be positive from an environmental point of view.

According to Lindén (1994), there is always a difference between an individual's thinking and doing regarding environmentally related actions. In order to relate attitudes to concrete actions, it is not enough to mediate knowledge and information. Factors that are of greater importance are, according to Lindén (1994), that the individual receives quick and clear feedback for his action, that he can control the results of his action, and that potential practical obstacles for performing the action are removed. Thus, according to this theory, reaching high on-site separation ratios in multifamily dwelling areas is challenging, as individual feedback is difficult, and the risk of demotivation, when individuals see that others do not participate, is high. Lindéns theory suggests that it might be difficult to change behavior through change of attitude. From this aspect, reducing practical obstacles becomes even more important, as the easier it is to do the right thing, the less people have to care, but, anyway, they behave in a way that is in line with environmental procurement. So, instead of asking ourselves how we can make people care, we might ask ourselves what we can do in order to create systems where it does not matter if they care or not, or that appeal to parameters other than environmental engagement. One such parameter is the longing to have a sense of belonging to a group. This can be illustrated by the example of transparent wastebins. These were initially introduced in order to decrease the risk of bombs being planted in public spaces. However, transparent waste bags can also be one way of increasing on-site segregation separation, through appealing to people's longing to have a sense of belonging to a group; if it becomes obvious what most people do, you do not want to exclude yourself from the majority and do different. In order to use this, solid waste management must become increasingly visual in the urban environment.

7.3 Development of Waste Collection Infrastructure

Several previous studies on the topic of waste sorting behavior state that convenience is one of the most important factors in increasing household waste separation (Ando and Gosselin 2005). Thus, further development of waste collection infrastructure will continue to be of key importance in relation to sustainable solid waste management. The aim is to develop systems where waste sorting becomes easy and a part of the daily routine for households, irrespective of their environmental engagement. This is done by increased understanding of the sorting in the daily routine and aspects which are of importance to the households.

Interviews performed in Malmö over the last few years show that households commonly dispose of household waste on their way out, rather than making a separate visit to the place for waste disposal. Thus, in many cases, entering the recycling building in order to dispose of waste in different fractions can be cumbersome for several different reasons, for example if the person in question is using a bicycle, baby-carriage, or similar that they would need to leave outside the waste disposal area; if residents have separated recyclables in special vessels, which need to be returned to the apartment or if residents feel that they get dirty from the waste disposal.

Therefore, a new design for waste sorting facilities is now being developed by the municipal facility owner, MKB. The main change is that waste will be disposed of in different fractions from the outside the building (Fig. 7.1). As each waste inlet door is opened only with an electric tag, disposal of waste from by-passers is avoided. The building is also equipped with scales under each wastebin inside the building. Householders get direct feedback from their behavior. The electric tag also makes it possible to track the waste disposal back to the individual household and, for example, generate monthly statistics on the amount of waste disposed of in different categories.

Fig. 7.1 Sketch of new recycling building under development in the Augustenborg area. Illustration Jaenecke Arkitekter

7.3.1 Issues to Address in Design of Recycling Buildings

Safety—Having only one entrance to a recycling building can result in a sensation of being unsafe leading to worry about potential disturbance or harm caused by intruders. It could be advisable to design the buildings according to a "pass-through" concept, that is, with two entrances so that residents feel less worried about being trapped inside by strangers.

Accessibility—24-h accessibility is needed to fulfill the needs of modern citizens with varying working hours and habits. Not providing 24-h disposal possibilities for all types of household waste—including special waste, such as bulky items—increases the risk of unwanted dumping of waste in places such as attics, basements, recycling buildings, or outdoors in residential areas.

Cleanliness—Lack of neatness often generates a decreased willingness to behave in accordance with the objectives of the waste management department. If the recycling building is neat and tidy, households are more likely not to leave them messy. It is important for households not get the feeling that it does not matter what they do, since others might not care, and so the effort made by some would be in vain. Thus, regular maintenance, in many cases this means daily, on the part of the facility manager might be needed to maintain high confidence and willingness for householders to make an extra effort.

Clear information—An increasing ethnical diversity calls for increased care in relation to use of symbols and pictures when communicating sustainable waste management to households. Classical errors must be avoided, such as use of before and after pictures in a left-to-right order, which is not compatible with the customs of Arabic-speaking citizens.

Placement of bins—Make it easy to do right and difficult to do wrong. A simple change of the location of bins for separately collected food waste from very close to the entrance to the building to further in results in a clear decrease in missorting; mainly through a decreased number of misplaced bags of residual waste in food wastebins. Using electrical tags for opening food waste chutes in vacuum systems can have the same effect, as this reduces unintentional misplacing by householders in the area as well as by by-passers.

Aesthetics and visibility—The evaluation of the systems for food waste collection applied in Bo01 by Åberg (2004) clearly state that behavioral changes are influenced by many factors other than planned information on the part of the municipality, such as communication with facility managers and visual impressions related to the disposal and collection of food waste. Thus, the disposal and collection structure also has an important role as a reminder of how households can contribute to sustainable waste management.

Flexibility—New fractions might become significant for on-site separation in the future. Thus, areas for waste disposal should allow for a certain flexibility and inclusion of new fractions.

7.4 Economy: Getting Paid for Resources

"From waste to resources" is a commonly used saying, but how do we get there in dollars and dimes? The concept of on-site separation of several different waste fractions—including all packaging fractions and newspapers under the producer responsibility legislation—was suggested as the norm on a national level in the Swedish Government Official Report on future organization of solid waste management in the country, presented in the summer 2012 (Swedish Government 2012). The same report suggested a relocation of the currently private responsibility for waste under the producer responsibility to the municipality. This would change the economic reality for municipal waste management, as charges currently paid by producers of packaging materials would be transferred to the municipal waste management budget.

The demonstration of a close link between waste management and energy production has resulted in an increased collaboration between the Municipal Waste Management Company in Malmö and locally active energy companies in the "Climate Contract" in Hyllie. The interest in biogas as a part of a sustainable energy matrix can, in time, result in increased value for separately collected food waste as substrate for biogas production. Several biogas production plants in Sweden are already experiencing an increased demand for food waste as substrate. Due to the need for pretreatment prior to digestion, it has previously been stated that biogas producers, in the very near future, will probably also charge for the treatment of this substrate (Swedish Waste Management Association 2011). However, new technologies are currently being introduced on to the Swedish market, minimizing the pretreatment through use of dry anaerobic digestion

(VMAB 2013). An increased use of this type of technology could increase the market value of household food waste and result in negative treatment costs in the future.

Another challenge in relation to sustainable solid waste management is the current lack of markets for digestate from anaerobic digestion of organic waste. A rather recent development of a certification scheme for digestate from biogas production plants receiving waste from urban, industrial, and agricultural areas (excluding wastewater treatment plant sludge) (SPRC, 120 2013) has increased the confidence in using digestate as a substitute for chemical fertilizers or undigested manure. Several scientific publications have also stated an increased fertilization value for digestate relative to chemical fertilizers, compost, and undigested manure (Lantz et al. 2009; Swedish Waste Management Association 2011). However, some actors in the food production chain, as well as some environmental organizations, are still reluctant toward using digestate as fertilizer (Arla 2008), probably much related to the debate regarding the use of sewage sludge on farmland, and, primarily, the risk of cadmium contamination as a consequence. In recent years, several key actors on the Swedish food production market have however changed their position in relation to digestate, which has been beneficial for the biogas production area. It is of key importance to maintain a high quality standard for produced digestate in order to maintain the confidence gained. Future threats may consist of increased contamination by heavy metals through anaerobic digestion of imported food produced with chemical fertilizers with high cadmium content, or contamination by microplastics as a result of physical pretreatment of packaged food or mis-sorting in household food waste, prior to anaerobic digestion.

Another complexity in relation to digestate is that the amount and relation between different nutrients in digestate can vary, both over the year from a specific plant and between different plants, depending on the input. Thus, it is difficult for the agricultural sector to compare the fertilization value in digestate with that of chemical fertilizers. Due to the main focus on wet digestion technologies in Sweden, the digestate produced commonly has a low dry matter content, which can make transportation inefficient.

Increased commercialization of digestate from biogas plants could potentially increase the value of the output, which could have an effect all the way back to investments made in providing systems for separate collection of food waste from the generators.

Reinforcement of recycling behavior through differentiated taxes for separated versus mixed household waste has a long tradition in Sweden through the producer responsibility legislation on packaging materials and newspaper. However, this has become increasingly interesting in relation to food waste separation in later years. Many municipalities, including Malmö, have chosen to decrease costs for collection and treatment of separately collected food waste, and increase them for residual waste. This is a clear signal to households, and it follows a "polluter pays" principle. However, a balance must be maintained in order to decrease risks of unwanted dumping of residual waste. Also, economic sustainability must be maintained when the actual costs for waste management are being covered, and also when households improve their waste sorting behavior.

In order to decrease the risk of intentional "mis-sorting" of residual waste as food waste in order to decrease collection and treatment costs, a mis-sorting charge was introduced in 2013. This was again a clear signal to households on the part of the policy-makers regarding the importance of maintaining responsible behavior in relation to solid waste management.

7.5 Sustainable Structures and Forms for Collaboration

Developments within the solid waste management sector in Malmö can, to a great extent, be related to the engagement of individuals within different municipal departments. Although this engagement is crucial, it results in vulnerability, as projects and processes tend to halt if the person driving it, for one reason or another, suddenly disappears. It can also lead to situations where knowledge and experiences gained during a project do not get transferred to the wider organization, and the wheel needs to be re-invented over and over again. Thus, there is a challenge in developing structures in which the development processes become less vulnerable to the presence/absence of a few individuals. However, in the initial part of a development process, the creativity and freedom of these "internal entrepreneurs" are essential. Thus, the institutionalization of the processes previously led by engaged individuals must be done at the right time in order not to interrupt a creative development process.

Another challenge relates to the externalization of solid waste management services. As an example, in Malmö, two different entrepreneurs have been contracted for household waste collection. Contracts are renewed every 5 years. As collection staff usually is the ones in direct contact with households, it is of great importance that these people are engaged and well informed in order to answer potential questions from households. One way to work on this is use of mandatory courses for all staff members involved in contracts with entrepreneurs. However, this issue is further complicated due to the producer responsibility legislation on packaging materials and newspaper. In areas with on-site separation, collection of such materials is generally performed by yet another or more entrepreneurs. These collection services are contracted by the national association of packaging and newspaper producers (FTI), and the setting of standards in relation to the use of renewable transportation fuels or education of staff is beyond the control of the municipality.

Due to the many times large investments are needed for the construction of solid waste treatment infrastructure, it is not uncommon to have long contracts between municipalities and treatment plants. Such long-term contracts might be needed in order to viabilize investments in waste treatment capacity. They also result in an inherent inertia in the system. Thus, although it might be demonstrated that treatment alternatives are more sustainable than the ones currently used, a change might be difficult due to these long-term arrangements. One way to handle such situations may be through close collaboration with the treatment facilities, as has been the case in Malmö. Here, development projects showing the viability of separate collection of household food waste were an important input to the contracted treatment facilities' choice of construction of a new plant for anaerobic digestion of food waste in the city.

Another organizational challenge emerges when waste management is connected to contracts with long-time horizons, such as the Hyllie Climate Contract. It is obvious that many or the public servants taking part in the process of developing the contract and thus establishing system boundaries and forms for evaluation and feedback, not will follow the whole process until the end of the contract period. Thus, internal communication and firm establishment of the strategies undertaken in such long-time horizon project are of high importance for the process.

7.6 Including Waste Management in the Urban Development Processes

As seen in the examples presented in this book, waste management has, in many cases, been seen as an "add-on" in an urban planning process which, to a great extent, already has been finalized.

Making other parts of the municipality with key influences on urban development processes care about and feel responsible for an efficient sustainable solid waste management, is still a challenge. The aim in this process must be to make waste management equally as important as transportation structures and buildings in urban development projects. At the same time, actors primarily responsible for solid waste management must understand the importance of integrating this into the urban environment in a way that is attractive from several aspects of importance within a sustainable urban development context.

Examples presented in this book state that inclusion of all relevant actors early in the urbanization project is of key importance. Thus, the development of decision-making structures and collaboration forms where key actors from the area of solid waste management are represented, as well as including waste management in binding contracts between landowners and developers, can be of key importance. Malmö has taken important steps in this direction in later years, but full recognition of the importance of acknowledging solid waste management in the urban planning process remains a challenge. Such recognition would also mean internal enforcement within the waste management department, in order to allocate more resources (both in terms of competence and time) to strategic issues. This would increase the possibilities for the department to play an active role in the urban planning process.

A strategy for increased recognition of waste management in the planning process currently tested by Malmö is to further increase the visibility of waste in the urban environment. While doing so, waste collection structures must also become increasingly attractive from an aesthetic point of view. Steps in this direction have, to some extent, been taken by the municipal facility agency, MKB, which, since several years ago, has had a policy to cover all their recycling buildings with green roofs. However, much more can be done in this area. Making waste disposal facilities interesting from an architectural point of view is regarded as an increasingly important feature in a sustainable waste management strategy.

The concept of maintaining solid waste visible in the urban environment, while at the same time also making the cities denser in order to avoid urban sprawl, also calls for challenging the conventional collection structure for household waste, for example, through introduction of alternatives to heavy collection vehicles in densely urbanized areas.

References

Ando A, Gosselin A (2005) Recycling in multifamily dwellings: does convenience matter? Econ Inq 43(2):426–438

Arla (2008) Kvalitetsprogrammet Arlagården/Quality program the Arla farm (in Danish). http://www.arla.dk/DAimages/Livet%20p%C3%A5%20g%C3%A5rden/V5-1424-Arla%20kval. prog.%20rev.dec.08-DK-Arla.pdf

Borgstede C, Biel A (2002) Pro-environmental behaviour: situational barriers and concern for the good at stake. Göteborg psychological reports, vol 32, No 1. Göteborg University, Department of Psychology, Sweden

Lantz M, Ekman A, Börjesson P (2009) Systemoptimerad produktion av fordonsgas—En miljö—och energioptimerad studie av Söderåsens biogasanläggnin/Systems optimized production of vehicle gas—an environmental and energy assessment of the Söderåsen biogas production plant. Report 69. Envionmental and Energy Systems Studies, Lund University, Lund Sweden (in Swedish)

Lindén A-L (1994) Människa och Miljö (Man and Enviromnent). Carlsson Publisher, Stockholm

SPRC, 120 (2013) Certification rules for digestate from biowaste by the quality assurance system of Swedish waste management. SP, Borås. http://www.sp.se/sv/units/certification/product/Documents/SPCR/SPCR120.pdf

Swedish Government (2012) Statens Offentlia Utredningar (SOU) 2012:56. Mot det hållbara samhället—resurseffektiv avfallshantering (in Swedish)

Swedish Waste Management Association (2011) Substratmarknadsanalys: sammanställning och analys av substratmarknaden/substrate market analysis: summary and analysis of the substrate market (in Swedish). Report 2011:23, Swedish Waste Management Association, Malmö

VMAB (2013) Västblekinge Miljö AB. http://vmab.se/vmab/biogasanlaggning/. Accessed 22 Mar 2013

Åberg H (2004) Boendeperspektiv på hushållsavfall och på system för insamling och behandling i Västra Hamnen, Malmö. Report 37. Department of Home Economics, University of Gothenburg, ISSN 1403-7033

The Way Forward

Keywords Challenges · Development

This book describes how several policies and objectives created on national and supranational levels in later years have been created in order to decrease the environmental burdens related to solid waste management. The operationalization of these policies and objectives is made on a local level, and the book presents such processes, using Malmö as a case in point. Several challenges in the endeavor toward sustainable solid waste management still remain and have been discussed. However, the way forward is clear:

- Creating systems for waste collection where it is cheaper and easier to act environmentally friendly than the opposite.
- Presenting attractive alternatives or participation in waste recycling activities in collaboration with external actors in order to also increase accessibility and user-friendliness in the actual household.
- Present flexibility in different solutions for waste separation and collection for different areas.
- Increase the visibility of waste management in the daily life of citizens, and, in this process, also increase aesthetic values related to waste management and decrease disturbing factors, such as smell and sound pollution.
- Work strategically with solid waste management and integrate this question into the wider urban planning processes in the city.

Continue collaboration with external actors in research and development projects in order to develop and evaluate new technologies as well as information/communication strategies, acknowledging that neither technology nor information alone can result in sustainable solutions.

A. Bernstad Saraiva Schott et al., *Modern Solid Waste Management in Practice*,
SpringerBriefs in Applied Sciences and Technology, DOI: 10.1007/978-1-4471-6263-6,
© The Author(s) 2013